The Cosmic Eggg, AKA The Primeval Germ

A Journey of 59 + 21 Zeroes

by

Richard Bruce Wallace

To my Grandfather George Wallace, who imparted to me at a very young age the word "Oblivion" and to the "Age of Aquarius."

"The Shen of The Ages"

Zodiac indicator of the great 2600 years comprising an age/ great

BOOKSIDE Press

BookSide Press
877-741-8091
www.booksidepress.com
orders@booksidepress.com

The Universe

A Journey of 59 + 21 Zeroes

59,000,000,000,000,000,000,000,000

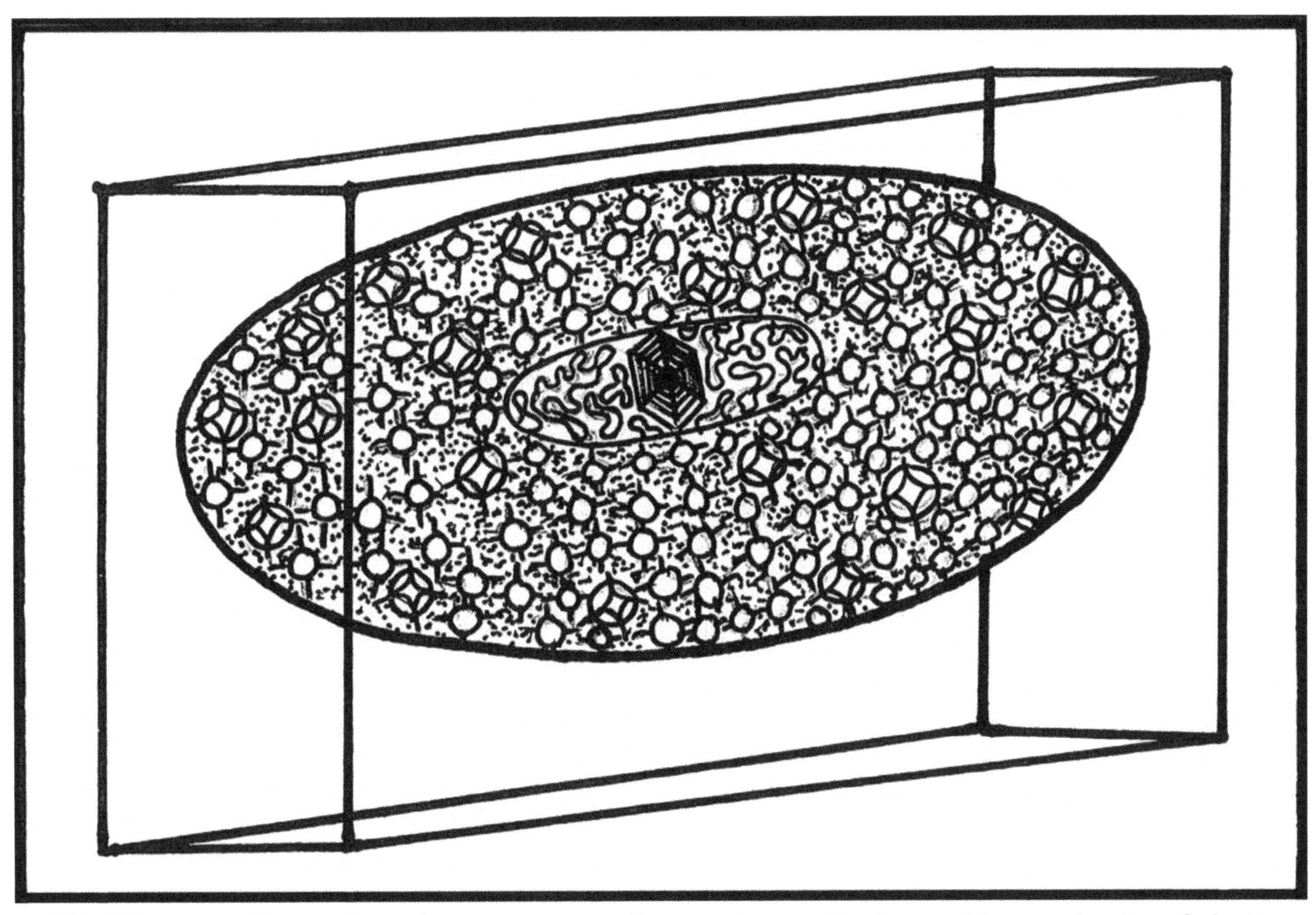

This Illustration, depicts, shape, distance concepts of inner and outer, allowing partial comprehension of what is external=light what is Internal=darkness/blackholes, that indicate that which is eternal

Contents

About the Author

This book is the result of a lifetime of encounters with many people who have been inspired. It begins for me at the age of four with a "nanny" whose name was Mrs. Rose. She assisted me with "understanding" the alphabet. My paternal grandfather George Wallace, who was a scholar in my eyes, and who taught me to speak French (against the will of my father) was my greatest source of self-inspiration. So at the very tender age of five I was on my way to understanding that there is an "order" to the world and that it is filled with different countries and people of different colors. My, father Wendell, who was a great provider (and my champion), insisted that I seek an education. In my own way and at my own pace, this I have done.

It was instilled within me then, to be the best that I could possibly, become. My grade school teachers, beginning with Mrs. Lynch, my third-and fourth-grade instructor whose specialty was the French language, and Mrs. Quinn, my fifth-and sixth-grade, English instructor, who introduced me to the "art of writing." There was Mr. Yates, who made such a profound impression on me while in the seventh and eight grades. There was Mr. Washington, who was able to impress upon me the understanding of mathematics; and Mr. Kollner, my history teacher. They, in their own ways, prepared me as I grew and developed a more complete understanding of the world.

In November 1965, at 21, I was inducted into the U. S. Army and six months later I found myself in Vietnam. From Fort Dix New Jersey, for basic, to Fort Lee, Virginia, for quarter master training I would travel. While stationed in Virginia I had the opportunity of spending time with one of my favorite aunts. Her name was Hazel, and before embarking on this journey she said to me, "Richard, go out into the world and find the "Light." It was there in this Asian country that borders the (green) South China Sea that I came to meet a Buddhist monk who became the inspiration of all my works that deal with religion.

One day while in the city of Saigon I encountered a group of white soldiers who were harassing him. These were regular white racist soldiers, doing what it is that they do. I could not allow this to continue in my presence and proceeded to fend off these small-minded individuals. This encounter was predestined and in the very short time of our acquaintance he instructed and demonstrated to me the power of religion. Some time later he would burn himself in a demonstration to his country men, "emphasizing" the meaning of their own beliefs. It was from him that I first heard these words: "All religion began in Africa."

Upon my return State side I was infused with dissention and disdain for the land that gave me sustenance. My life before Vietnam and afterwards has been a battle with these "Racists." I was moved to put down my pistol and rifle and I picked up the pen. My first book was called *Black Thoughts* (1968), a collection of poetry that dealt with this subject of "Racism." In 1983, I wrote *The Old Religion of Ancient Egypt*, the foundation on which this text was established.

Lastly I would mention Mrs. Wilson, who mothered my desire of putting ideas into print, and my own very dear maternal mother Margaret V Hawkins Wallace, who made life possible for me.

This book is the complete story of the creation of the universe, as it was understood by the ancient Egyptians. It is a collection of harmonic and radical "Black Thoughts" and the pursuit of equality for all of this planet's inhabitants.

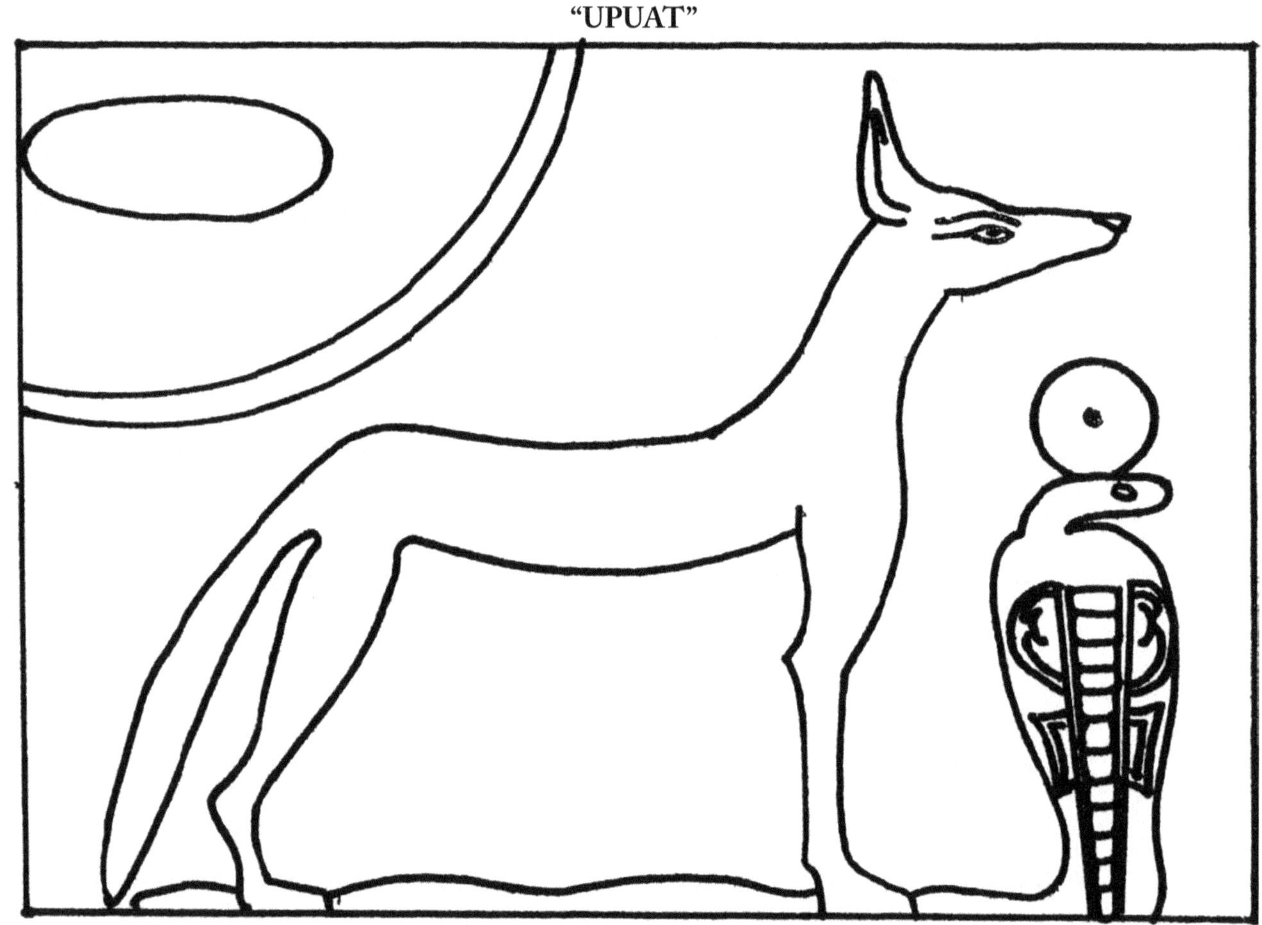

The "Opener of Ways" and the Solar Serpent "MAYHEM" announce the rediscovery of the "Cosmic Egg"

Introduction

Many are the mysteries concerning the universe and the life that is found within it. This is also true regarding the theories (laws) that govern it. In this book we examine the most popular ones, past and present, in search of the common constant that binds life, laws, and the universe. It exposes the views of the early hominids of Africa and those of the ancient Egyptians. It includes the Assyrians, Babylonians, Chaldean and ancient Sumerians of Asia, up to the Europeans and modern, day Americans. By combining the knowledge of all of these cultures and that of mankind worldwide, a stunning conclusion is reached. The concept of "Unification" is confirmed and it is realized in "The Order of Existence."

This text reveals the progressions of humankind, beginning with the need to seek shelter, up to the mastering of fire, the development of science, and the need for religion. The idea of "gravity" is reviewed using concepts that were unknown to Isaac Newton. The speed of "light" contrary to Albert Einstein, is not the fastest known phenomenon in existence. The "atom" is decimated and reduced to its "quantum" state and then into "strands" and strings and finally into the realization of oblivion. From gamma blast to black holes, supernovas to the big bang and to new age concepts such as the in-nerverse and the outerverse (including the identification of intrafarance) these ideas are dissected, reviewed, and then represented.

This book is truly "New Age" and of course this new age is "Aquarius." It is critical of many great works and acclaimed figures whose findings predate the recent discoveries of the United States space program (N.A.S.A). Its usage of the super telescopes which only recently (1989) have provided us new information in this age-old quest for understanding the mysteries of the universe, makes many of these theories obsolete, belonging to "ages" long past. C.O.B.E. or the Cosmic Observatory Background Explorer and the rediscovery of the cosmic background radiation (or the "Cosmic Egg") have shed new light on how the universe will be viewed in the future.

This work is written in a style that allows the beginner to begin understanding how our universe works, while also providing new insight and food for thought for the "learned Ones." It examines in great depth the origin of the "planet" and its life forms, going on to include the origin of our sun (RA), the stars in outer space, and the creation of the "Light" itself. The text is developed in a manner that induces partial comprehension of the uncomprehendible, phenomena and anomalies that are found within the universe. Through out the past "ages" humankind has taken for granted the light of our sun, not realizing that "Light" is a universal phenomena. When C.O.B.E. downloaded its data and its spectrographs produced images of this so-called "Afterglow" of creation (the cosmic background radiation), the largest "Anomaly" in the universe was revailed.

The universe is made smaller with the understanding that this cosmic background radiation requires two-thirds of the existing space within the universe itself. That is a distance of the number 59 followed by twenty-one zeroes or ten billion light years. If this is the "Afterglow" of "Creation" and "It" has an absence of "light" it should suggest that the "light" exists in the remaining one third. Thus we conceive "New Age" concepts that are supported by the recent findings of modern-day space technology and the "Super telescopes." In this manner we advance science, by deducing that there must be an "Innerverse" of leftover "Afterglow" (that is now recognized as the cosmic background radiation) and an "Outerverse" where the "Light" is found.

The appearance of Earth life forms and the order of their appearance over millions of years demands the appreciation of advanced preparation, solidifying the concept that earth life forms are not here by chance or accident. Only in the understanding that there is an order of existence for the earth and its life forms do we prepare ourselves to comprehend the order of the universe. Can we gain the knowledge of the stars without the understanding of our own star (sun) RA? I think not. For the expertise on this subject I of course related to the ancient Egyptians and their so-called pagan views and was astounded to find clues that suggested to me they were aware of outer space. The thousands of years that these pagans spent worshipping the Sun produced a remarkably large volume of texts that are not to be found in the Hebrew or European cultures.

Did Isaac Newton really discover gravity? Was Albert Einstein correct in assuming that light is the fastest-moving phenomenon? How do the findings of Stephen Hawkins and his research on "black holes" give new definition to the concept of "oblivion"? Was there really a "Big Bang"? In this text we deal with it all, including the newest concepts of quantum mechanics and the String theory with its dimensional additives. Ausaures, the first "High Priest" of the Sun, said that "in the whole of the universe nothing is ever added and nothing is ever lost."

The abbreviated history in this text is factual, revealing the understanding of long-past cultures as it pertains to the understanding of the "Stars." Surprisingly, it seems that cultures that lack this enlightenment are more barbaric (racist), and history bears witness to this fact, prompting this author to ponder this very important question: Is it the pursuit of knowledge or the lack of it, the root of earthly racism? Where did it and how did it begin? The reader becomes the judge and the jury as this historical accounting unfolds, spanning the millions and millions of years that hominids have existed here on earth. It is thought that every homo sapien is a hominid but that not every hominid is a homo sapien. What hominids and homo sapiens have in common is the earth and the observation of the stars. Both their needs were the same: food, shelter, clothing, etc. Only technology and millions of years separate the two. They, like us of today, were Earthlings and oddly enough, the origin of God starts with them.

Lastly, this book is designed to uplift the nations of black peoples everywhere and for those who were stolen away from their homes and sold into captivity in America. It is meant to realign our senses with that which was taken away from us by force, not just our native languages but mainly our collective "Destinies" that at one period of times past were interwoven with one another. That is one destiny for all blacks and all Earthlings. It recaptures our past glories and re-presents itself in our understanding of the stars, reminding us that there is nothing inferior about being black.

In this new Age of Aquarius (the age of space exploration), the one must stand for the All and the All for the One. For there is no fulfillment that is greater than the understanding of the stars and the light that they produce.

Prelude: Concerning the Ancient Egyptian Religion

The Greek historian Herodotus affirms that the ancient Egyptians were "beyond measure scrupulous in all matters appertaining to religion." He made this statement after personal observation of the care they displayed in the performance of religious ceremonies, the aim and object of which was to do honor to the gods, and of the obedience they showed to the behests of the priests who transmitted to them commands declared to be and accepted as authentic, revelations of the "Will" of the Gods.

To outside writers, beginning with the Greeks, we of the modern-day western world (America) owe our understanding of the ancient Egyptians. The great majority of these writers, like the Greeks, were Europeans (and) from their deluded interpretations we receive the misunderstandings that are perceived as the truth today.

Other ancient nations were content to believe that they had been brought into being by the power of their gods operating upon matter, but the ancient Egyptians believed that they were the issue of the "Great God" who created the universe and (that) they were of direct divine origin.

However, the Egyptians were not only renowned for their devotion to religious observances. They were famous as much for the variety as for the number of their gods. Animals, birds, fish, and reptiles were worshiped by them in all ages, but in addition to these, they adored the great powers of nature, as well as a large number of "Beings" with which they peopled the heavens, the air, the earth, the sky, the moon, the sun, and the stars.

Expressions that are the result of a series of beliefs in tree gods, desert gods, water gods, and gods with human passion abound, and it is "these" which have drawn down upon the Egyptians the contempt of the Hebrews, the Greeks, and the Romans, and even modern skilled investigators of the ancient Egyptian religion.

Gaston M. Maspero said: "When I began to study the religious texts, I found that they did not breathe out the profound wisdom which others had found.... Certainly," he says, "no one will accuse me of wishing to belittle the Egyptians; the more I familiarize myself with them the more I am persuaded that they were of the great nations of the human race and one of the most original and most creative, but (at the same time!) (that) they always remained half savage."

There are two ways to perceive the ancient Egyptian religion: Either it was "all-knowing" or it was not. If the reader is inclined to think that negative reports such as those mentioned have merit, they are advised to not continue reading. These outsiders and their "lies" will become the "truths," revealed, as will the duplicity that today is accepted as religion.

Clouds span as much as 10B light years

Wispy clouds of matter detected by a NASA spacecraft are the biggest structures ever found, scientists say.

The largest span two-thirds of the known universe, or 10 billion light years, which is 59 billion trillion miles.

That's 59 followed by 21 zeros. It's also 120 million billion round trips between the Earth and moon.

The smallest span 500 million light years, about 2.9 billion trillion miles. —AP

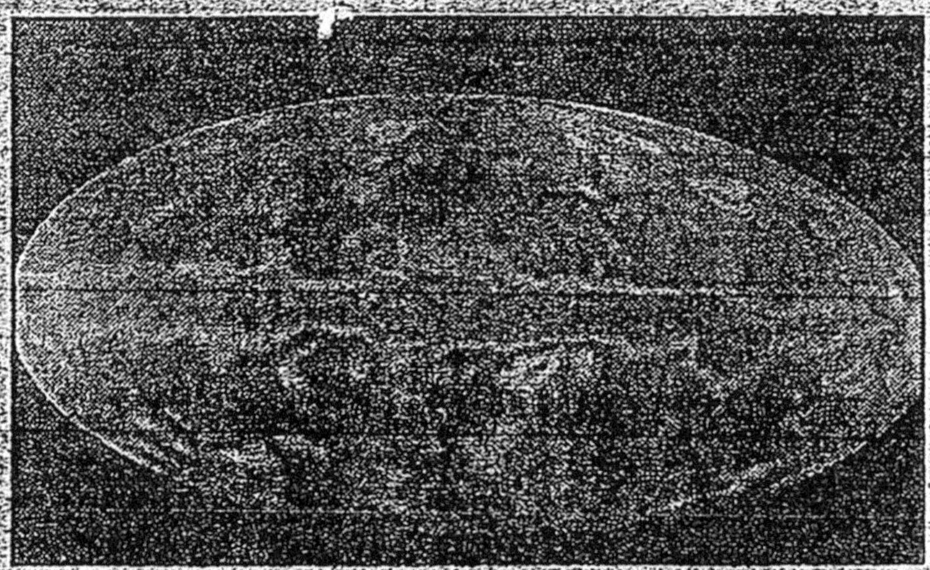

WHAT REMAINS: This map of the universe was produced with data from a NASA satellite. Patchy areas top and bottom represent ripples of wispy matter left over from the 'big bang.'

AP photo

Preface

The Cosmic Egg, theory or fact? The Cosmic Egg is a geometrically shaped anomaly in which the very ancient Egyptians envisioned the great mystery. It is a story that involves "the everything of matter." When this subject was rediscovered by French and German scholars, written in the vernacular of ancient Egyptian symbols and characters, its true interpretation was lost.

In the old days (before space exploration) scholars from around the world flocked into Egypt in search of its secret past. They all found it difficult when retranslating this amazing theme. For at that time there was no way to confirm or deny positively this "cosmic saga" that comes to us out of ancient Africa.

In the educated minds of these European savants (of those days) it seemed wholly doubtful and highly unlikely to have any relationship with the factual truth. After all, science as we understand it today was just then entering its infancy. Once again I restate the purpose of this book, which is to prove them wrong and to show that rejecting this, the most [Hakka] sacred of all the texts found within the tombs or temples of ancient Egypt, is a starting point of error and that the science and religious doctrines-then and now-lacked the meaning of its true retranslating. And to show that real science was first used in ancient Egypt and how it was developed out of the awareness of their religion that celebrated the birth of (light) the sun. (RA).

Using the Cosmic Egg as our point of initial focus, this story begins. The drawings presented here were developed by myself, with the aid of my younger brother Dana Ptah Williams. The photographs were captured by the United States Space program, N.A.S.A.

"Slide 28": Following subtraction of the dipole anisotropy and components of the detected emission arising from dust (thermal emission),hot gas (free-free emission), and charged particles interacting with magnetic fields (synchrotron emission) in the Milky Way Galaxy, the cosmic microwave background (CMB) anisotropy can be seen. CMB anisotropy—tiny fluctuations in the sky brightness at a level of a part in one hundred thousand—was first detected by the COBE DMR instrument. The CMB radiation is a remnant of the Big Bang, and the fluctuations are the imprint of density contrast in the early universe (see slide 24 caption). This image represents the anisotropy detected in data collected during the first two years of DMR operation. Ultimately the DMR was operated for four years. See slide 19 caption for information about map smoothing and projection.

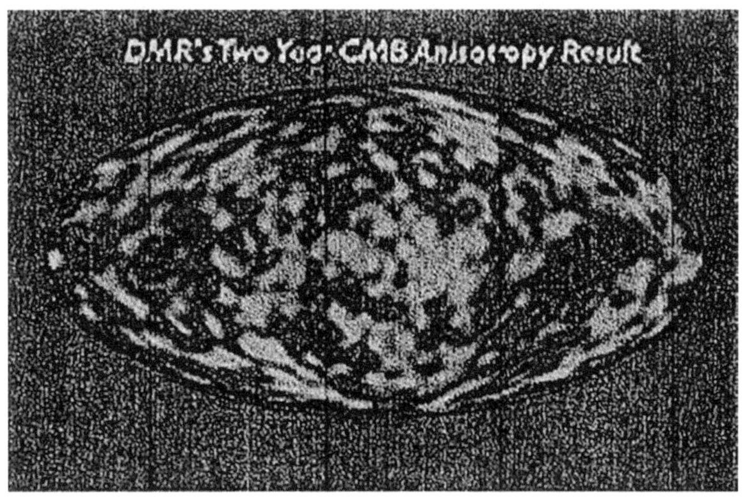

COBE's Orbit

The COBE Satellite

Early Development
of the Universe
Big Bang
Big Bang Plus Tiniest
Fraction of a Second
COBE
Big Bang Plus
300,000 Years
Big Bang Plus
15 Billion Years

DIRBE 25, 60, 100 μm Composite

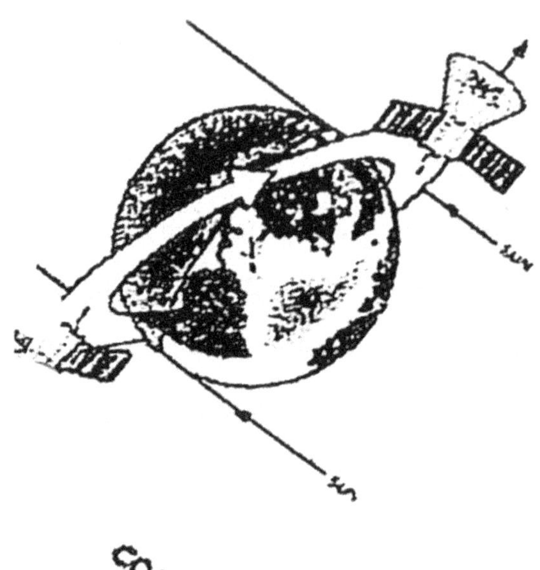

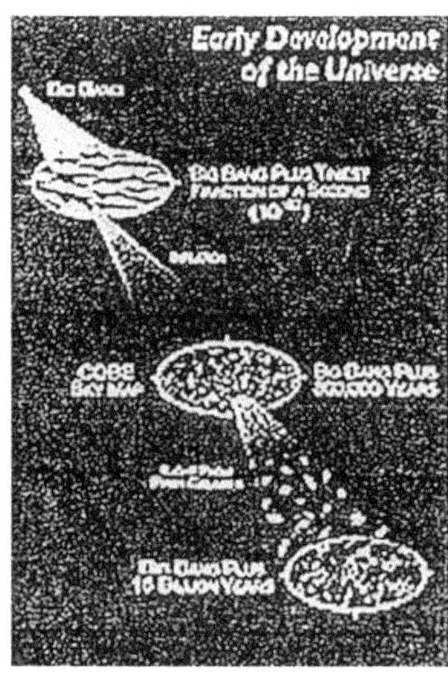

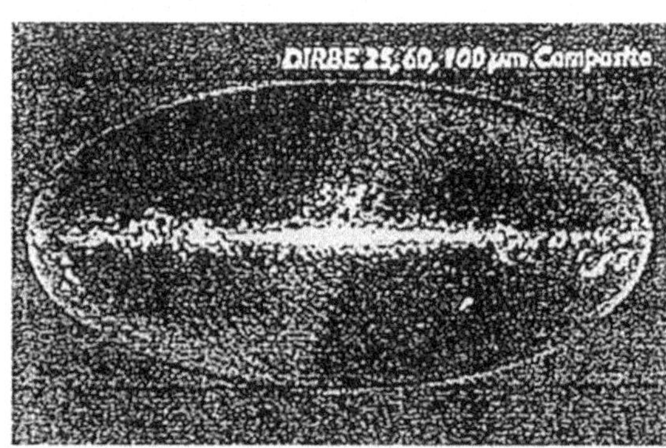

Foreword

C.O.B.E. or the Cosmic Observatory Background Explorer was launched on November 18, 1989. This satellite was designed to measure the diffuse infrared and microwave radiation from the early universe. The spacecraft was developed by N.A.S.A.'s Goddard space flight center. It carried on board three extremely high-tech instruments: the Far Infrared Absolute Spectrophotometer (F.I.R.A.S.), which made a precise measurement of the spectrum of the cosmic microwave background; the Differential Microwave Radiometers (D.M.R.), which detected for the first time and was used to characterize faint fluctuations in the cosmic microwave background corresponding to density structure in the early universe; lastly, a Diffuse Infrared Background experiment (D.I.R.B.E.), which obtained data that can be used to seek the cosmic infrared background and study the structure of the Milky Way galaxy and the interstellar and inter planetary dust.

The craft's orbit nearly passes over the earth's poles at an altitude of about 560 miles. The orbital plane is inclined by 99 degrees to the Equator, causing the orbit to precess (turn) to follow the apparent motion of the sun (RA) relative to the Earth. The spin axis stays pointed almost perpendicular to the direction of the sun (RA) and in a generally outward direction from the earth. C.O.B.E. orbits the earth once every 103 minutes. It views a circle of the sky 94 degrees away from the sun (RA) and as the earth moves around over the course of the year the spacecraft gradually scans the entire visible sky. When C.O.B.E. began to download its data, information and pictures, it had rediscovered the Cosmic Egg.

The instruments on board seemed to suggest that this structure contains the earliest source of universal radiation, Subatomic particles, and microwaves that were possibly left over after the "Big Bang" or existed before the "Big Bang." An article in a Boston newspaper, April 1993, suggested that to observe this photograph of the (Cosmic Egg) or the "cosmic back ground radiation" is to look into the face of God. This was nearly the same conclusion that the ancient Egyptians (minus modern-day technology) had arrived at.

In 1885, Dr, Henrich Brugsh collected literature on ancient Egypt and published "Religion and Mythology der Alten Aegypt" in Leipzig, Germany. He states that "the primeval spirit (Urgiest) felt the desire for creative activity and that his words awoke the world to life, in a form in which it had been mirrored in his mind. And that the first act of creation began with the formation out of the primeval watery mass of an "Egg…."

The Cosmic Eggg

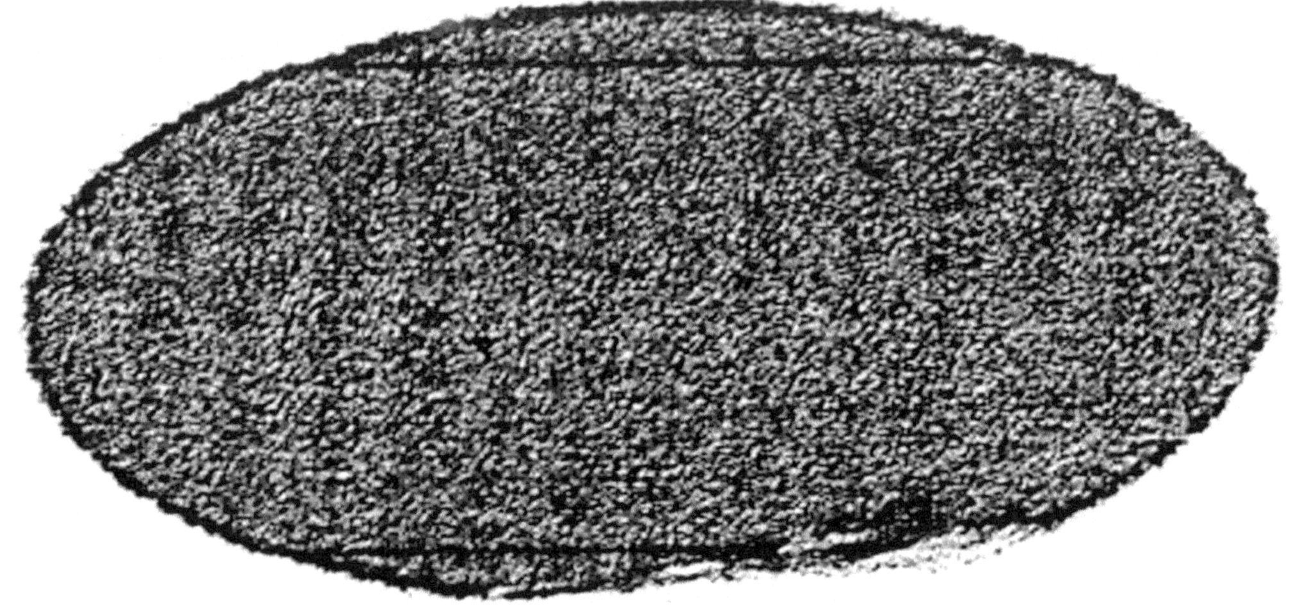

"THE COSMIC EGG" (a.k.a) "THE PRIMEVAL GERM"

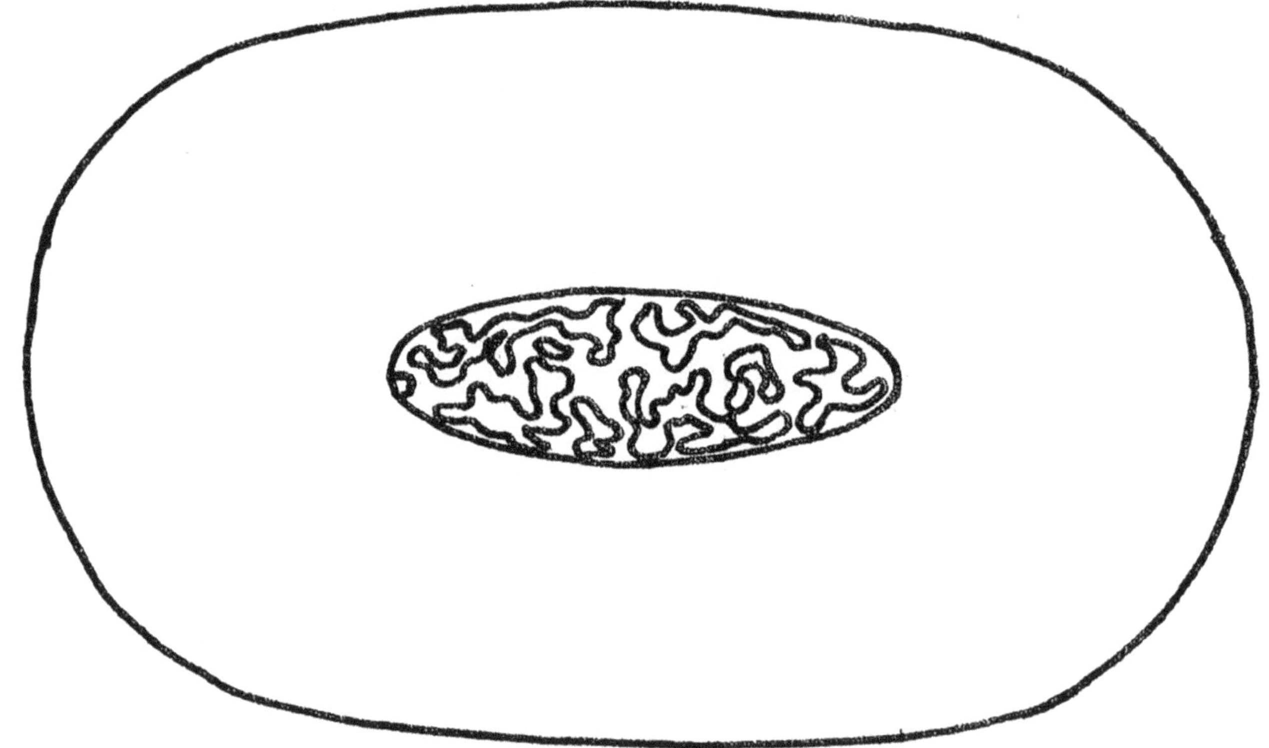

The Ancient Egyptians adhered to a theory that supports the "Space Germ" as the very beginning of evolutions.

"The Toughest Job In The World Is To Be A Man."

Chapter 1
The Cosmic Eggg

The ancient Egyptians inherited from their earliest ancestors a story of "Creation," in which they envisioned the beginning of life. This beginning involved the "Cosmic Egg," also known as the "Primeval Germ." It existed then, now, and always. They believed it to be the source of all things that are found within the universe.

All molecular matter, subatomic and atomic particles, owe existence to this Germ that found a haven within the Cosmic Egg (cosmic background radiation.) Their religion was derived from this system of cosmology.

Under this system, all the forces of the universe, along with the basic earthly elements, were associated with a super being or god or goddess. (Without this understanding, they would appear to be a culture of pure polytheistic practioners.) Every thought that could be reasoned was described and defined. Every force had a name and a number and was given a value. All comprehensible thoughts were written and logged on scrolls of papyrus paper. These facts were scrutinized many times and then complied into one book. This book was called the book of "Great Enlightenment." It contained all that they knew about the Universe and from this book we first hear mention of the "Cosmic Egg."

The process by which they gathered information, observed, experimented, took notes, and made conclusions seems very scientific to me. Is it possible that before the Chaldeans, Sumerians, Bablyonians, Assyrians, Persians, Greeks, Romans, and other European nations, that an African culture was able to attain such a high state of universal awareness?

The ancient Egyptian perception of the "Cosmic Egg" (cosmic background radiation) bear witness to this fact. How was this possible, that an African civilization that many believe to be remnants of an Ice Age culture could describe so vividly that which was unknown to the modern world until the spectrographs of C.O.B.E?

Walter A. Fairservis Jr. states in his book *The Ancient Kingdoms of the Nile* that there is archaeological evidence that life most likely existed on the shores of the Nile river, two million years ago. "According to ancient Egyptian texts, two million years ago the inhabitants of the Nile Valley believed in the "Creator of the Light" and an "Afterlife," there by establishing a religious theory that is perhaps older than mankind." So as we begin to realize that there was knowledge before the Greeks, understand now that this knowledge existed in Africa.

They possessed an intunement that was based on the natural order of the Universe and not on the logical deductions of the Greeks or the Romans. This "generic" understanding was the result of the natural order that great rivers obey. Of all the continents on the Earth, Africa has the mightiest of rivers. The ancient Egyptians had the Nile.

The Nile may be called the river that fathered and mothered Science. One reason for the rapid development of culture in very ancient Egypt is the natural fluctuations of this river. Africa and its many great rivers naturally prompted the development of religion, mathematics, and science. As prehistoric people built shelters and made their homes by these rivers, they experienced the natural inundations and became wiser.

This natural sequence of inundation awakened in the minds of these ancient Africans the need to measure distances. At other times of the year, these sources of water resided and dried up. During these periods, they learned to conserve water. The Greeks were fond of saying that "in Egypt, they store water in jars."

The ancient Egyptians conserved water. Thus, as the river went about its annual cycle, they began to measure and analyze these changes and acted accordingly. Through everyday survival practices such as these, science was born and the need for mathematics discovered. In this early era, they were actively involved in the observation of "the stars." Of course our sun (RA) was their initial point of focus.

The ancient Egyptians believed that the earth was once a part of the sun. "RA" was their name for the solar furnace that gives the earth life. In their language, RA was the first written and spoken word and was their emblem of the "Creator of the Light."

Today scientists acknowledge that we are the by product of "stardust." (In ancient Egypt that would not be surprising news.) With the aid of the great telescopes, astronomy and science make new discoveries almost daily. Upon careful review of their findings, they appear to be "rediscoveries" of this ancient African space tale and its connection to the whole of the universe, through the "primeval germ" and the "Cosmic Egg."

Scientists today are involved in a endless debate as to the origin of the earth. This was not an issue in ancient Egypt. Their name for the planet on which we live was "Seb." This name is suggestive of solar origin. Seb, like many of their names, reflects the dualism of the earth and the sun. (RA). When read left to right, Seb means a solid or the outer core of the planet. When read right to left, Bes is the word for fire as in the nuclear inner core of the earth. In their minds they related the two furnaces, one that is outwards and the other inner-both contain the same life-giving atomic structure of "heat" and "movement"-arriving at the conclusion that the earth was once a part of the sun.

Perception of seen and unseen forces, such as the tiniest microscopic particles, was a "constant," for meditation and always on their minds. The need to understand the "order of existence" was primary. Somehow they were aware of radiation, subatomic particles, and atomic particles and their relationship to molecular matter. Their creation theme speaks of a "Creator" that lived in an "Ocean of Darkness," before speaking his name and straight forth becoming ethereal matter. In their minds the "Darkness" gave birth to the "Light."

Today we acknowledge the findings of this African culture, but this is done in away thats lights and does not accredit its contributions, to science, mathematics, or religion. In my mind we (the world) are worse off because of this. We stumble and ponder our recent findings, rejecting this input that was so painstakingly collected, written on the walls of great temples and inscribed in the tombs of the long-ago dead. They, in their might and glory, existed within the same universe as do we; they were born to a cycle of life and death as are we. To disregard their findings because of their African heritage is not only fool hardy, but borders on the edge of ignorance and racism.

"SHU" The God of "GAS"

Shown with his feather which is the symbolic equivalent of his name. He carries the torch of light that is passed on from one "Age" to the next in his right hand and the "Rasaurian Ankh" in his left hand.

"TEFNUT"

The Goddes of "Liquid" above her is her name in Hieroglyphics

"SEB"

The God of "Solid" above is the "Sacred Goose" symbolic of his name.

"Our Lives Are Not Promised Tomorrow; We've Got To Love Each Other While We're Here."

Chapter 2
MATTER

The universe is made of particles. Some are atomic in weight and are easily recognized. Others are subatomic and possibly without weight. These were the seen and unseen forces that the ancient Egyptians called the results of "Hakka" or the "Self-Created One." The origin of "God" begins here, with this theory.

Hakka is the complete story of the how, when, where, and why the universe came into being and where it would go, if ever it came to an end. We must try and understand that the ancient Egyptians believed in a perpetual universe, never-ending. That is to say, in their minds a beginning automatically would suggest an ending. The end of any thing allows the beginning of another. Thus the beginning is an ending and the ending is a beginning. It allows for the conception of the Cosmic Egg and the primeval germ. It deals with the notion of life that begins in oblivion and ends in oblivion.

Ausaures, the first high priest of ancient Egypt, said that "in the whole of the universe nothing can be added and nothing can be lost." The primeval germ and the Cosmic Egg explain the evolution of radiation and subatomic particles, while their "Creator of the Light" explains how matter was evolved from these subatomic "radiant particles."

This concept was never clearly understood by the Greeks, Romans, and other outsiders that have poorly defined the wonders of the ancient Egyptian cosmology. Particles that possessed "heat," "weight," and "movement" they associated with ethereal matter that could be changed, rearranged, and altered. Subatomic particles were labeled as nonethereal matter, a form of universal "radiation" that is not "emitted" from stars, seemingly lacking recognizable heat emissions and atomic weight and unable to be changed or rearranged or altered.

To fully explain their understanding of nonethereal matter and ethereal matter involves a three-step process that begins on the earth (Seb), leads to the stars, and finally leads to the primeval germ and the Cosmic Egg. This first step will be accomplished by identifying with their religion and their hierarchy of gods and goddesses. They will escort us to the stars and from there, we will arrive at (the findings of the Cosmic Observatory Background Explorer) the cosmic back-ground radiation.

Shu, the first of the earth gods, was believed to be the identification of the unseen element "Gas." He is followed by a female counterpart, Tefnut, who represents the element "Water." Completing this trinity of deities is Seb, representative of the final element of ethereal matter, "Solid."

Upon careful examination, we find that this trinity (gods) of elements is indicative of scientific thought and was present in the minds of these early Egyptians before the concept of religion. It would seem that science was the inspiration for the beginning of religion. So today we can question the validity of science without religion and religion without science and trace them both back in time, establishing the creation of matter, the birth of science and religion, and the development of mathematics on the earth.

Shu was the right arm of RA (the spokesman of the "creator of the light," RA Nebertecher, who claimed to be the original source of the light and was "self-created," from subatomic particles within the darkness of outer space. They are the sacred trinity of the light.) He was stardust converted into "gas," the original earthly element. Tefnut was the "water" that sustains life and the visible transformation of matter. Seb is the "solidity" that accounts for the final altering of this "One" element that "Exists" as "three."

They represent the very first trinity of earth gods and goddesses, while the "sacred trinity" pertained to light and substances not of this earth. From this understanding of matter, science began to emerge. It is a natural presumption (not a logical one) to presume that these deities also represented the numbers "one," "two," and "three," respectively. And from the earliest of times, they are to be found "symbolically" and in the written text and in "all" aspects of their culture. This pentagon of deities represents the gods and goddesses of all matter.

The understanding of matter, then, now, and always, begins here. The earth was the domain of these "elements" and "deities" and was special, more so than any other place in the universe. If the reader is to comprehend their much more complicated belief in a "creator of the light" (matter), then understand (that) this was a "religious enlightenment," based on scientific data.

The story of this "creator" that comes from the ancient Egyptian text reads "I am 'He' that existed within the Primeval Ocean of Darkness, before the beginning of Time, which began when I spoke my Name, as a Word of Power." This is the story of "stars" and the "origin of the "light." It is a concept that separates that which is savage in humankind and awakens the pursuit of religion and science; the understanding of these factors is what produces an enlightened being.

The reader shall experience the how and why of this transformation from a savage hominid to Homo sapiens and finally to the realization of Man and why his cultures and civilizations occurred and were first developed in prehistoric Africa. All this begins with the observation of the stars, for there is no fulfillment to be found that comes close to the understanding of the stars and the light that they produce. The Greeks were of the opinion that asking the question: Who am I? was proof of their existence. The ancient Egyptians (hominids) looked upwards to the stars as their proof.

What once was a theme, to be ridiculed and instantly dismissed: the idea of a Cosmic Egg. A life form or germ that existed before the stars was pure insanity. Back in those days of Darkness, fact was understood as fiction. With the realization of the cosmic background radiation which we will equate to the cosmic egg we will find the source of the origin of the Light, and fiction now becomes fact.

In the primitive minds of these Africans, there was a primeval germ that existed before the Darkness and the Light. And It is responsible, for the coming into being of Itself, subatomic and atomic particles, and the entire universe, as we understand it today. This concept represents the understanding of another source of energy that is beyond the light, and this beginning was Radiation.

The identity of these three, Shu, Tefnut, and Seb, should not be confused, as did the scholars of the past when trying to separate these different facets that explain the forms of existence, and of all matter. They are the gods and goddess of Light (matter). Its mirrored reflection is Radiation, the source that explains the origin of the Light. As the reader will come to understand, one is the explanation of the other.

Some scholars of today claim that the ancient Egyptians were unaware of the existence of Antimatter, dark matter, and radiation. (How could they know of Radiation before Madame Curie?) They profess that without the knowledge of and the usage of the telescope, black holes, supernovas, and gamma explosions most certainly were unknown to them. And as without such understanding, it would be impossible for their doctrines to perceive completely the origin of the universe, the result would be their not knowing that, these are the main clues to solving the origin of all life forms within the universe.

So naturally, it was assumed that these black Africans could not and would not have had a valid clue to the understanding of God.

They did not have the advantage of modern technology and all its fine expensive equipment. But by adhering to the "order of existence," all of these concepts were known to them, as were other ideas involving the establishment of the universe that even today (before this text) are unknown, with all of our sophisticated machinery and instruments. Allow me to put this negative concept to bed and say that the "Great Pyramid" was an observatory and a telescope made of stone and it existed before the realization of optical glass.

Mentioning this I do not intent to draw away from this text, because for many the Great Pyramid is symbolic of Pharaonic Egypt. Let it suffice in the understanding, that the one piece of the puzzle missing in the search for the solution of the very beginning, that includes antimatter, dark matter, supernovas, blackholes, and Gamma Blast is the evidence that was discarded and ridiculed-the concept of self-creation.

As we shall come to understand, there was not that much, if anything at all, that was unknown to them concerning the makeup of the universe. The original laws of matter were known to them and, as this text will show, that scientific understanding was coded within their religion, which announces, "From the darkness came forth the light.

"OBLIVION TO OBLIVION PARTICLE CHART"

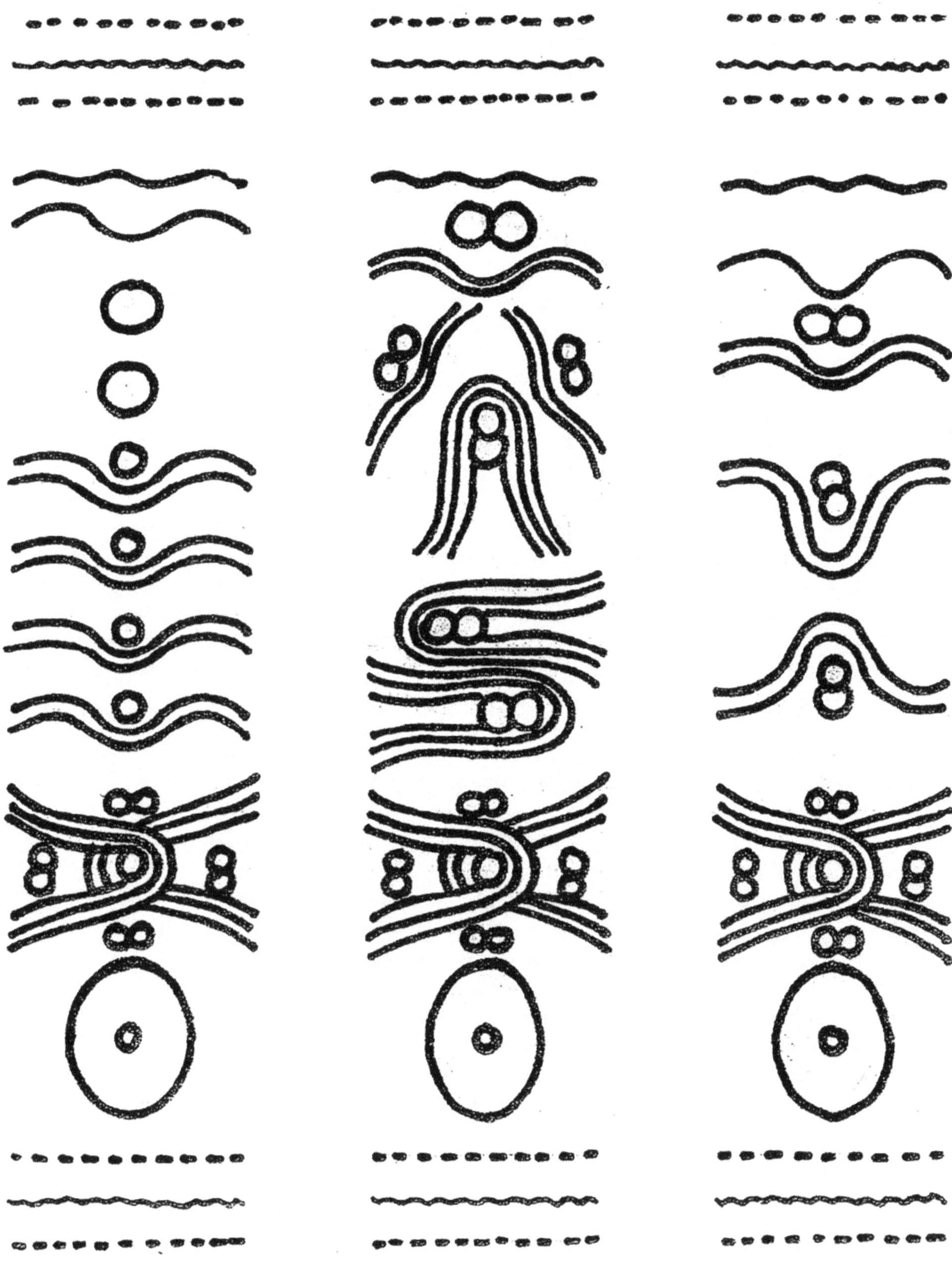

"RANEBERTECHER"

"THE LORD OF LIGHT"

"An Ear That Hears, Follows The Eye That Can See."

Chapter 3
The Light

One of the greatest mysteries, of all the ages of humanity, is the origin of the "Light." The How? When? Where? Why? And What is "It"? makes this so. Even today, the year 2005 A.D., these question still remain. It is true that we have a greater understanding of matter (and light is the equivalent of matter).

However, the origin of these photon particles is still a mystery. We can estimate the speed of light under normal earthly conditions-(186,000 miles per second, give or take a little). We're able to separate the colors by disturbing these photon particles as they pass through a prism or are reflected by water, or observe them as they are bent by particle debris, as they traverse the vastness of deep outer space. There is light that has heat and light that doesn't. Before man was able to produce light (Edison), insects and minerals in nature were able to duplicate this phenomenon. Finally we can say that light comes from stars.

How do we know the speed of light? It was an Italian named Galileo who, many think, was the first (European) man on earth to attempt to measure the speed of light. He was able to place two observers two miles apart on a very clear night. Each observer was given a lantern that could be exposed at the given command. This period between the uncovering of the lanterns was to be measured. He failed in this attempt.

Danish astronomer Ole Roemer is credited with the first breakthrough in 1676. He determined that light travels at 192,000 miles a second. Using his telescope while observing the planet Jupiter and the eclipses of its moons he noticed that the intervals between the eclipses of one moon were not the same, being 16 minutes, 26 seconds greater at one time of the year than at another. He was able to conclude that this difference could not be due to any real difference in the period of the eclipses but must be caused by the greater distance over which the light had to travel from Jupiter to the earth, when the earth was at different parts of its orbit and (therefore) nearer to or farther from Jupiter. Was this the reason that Galileo failed?

Perhaps the speed of light requires millions of miles of travel in order to be measured? Various methods of measuring the speed of light have been adopted since then and oddly enough they all give practically the same results.

Two French scientists, Foulcault and Fizeau, flashed beams of light between systems of mirrors and lenses ("using a method that is too difficult to describe") and concurred that the speed of light is nearly 192,000 miles per second. There are those that think that the best modern "astronomer" was an American, Albert A. Michelson. In 1924 he flashed light back and forth between various California mountain peaks over a distance of as much as 90 miles. He announced the speed of light to be 186,285 miles a second. His findings are assumed to be correct and today the speed of light is estimated at 186,000 miles per second.

Natural light that comes from our Sun (Ra) has heat and contains "radiation" in its ultraviolet rays. Light that is produced by fireflies, deep sea fish, or decaying tree trunks and logs that glow at night are considered cold light (lacking radiation). Light that is produced when a substance is heated to a high temperature is called incandescence, such as that of our electric light bulbs. There is neon light that is produced by adding an electric current to a glass-filled tube of this substance. Various gases, especially those called inert gases, are used for this purpose.

Some minerals shine brilliantly and emit beautiful colors when they are illuminated by certain kinds of invisible light such as ultraviolet rays. This kind of luminescence can last after the invisible light has been withdrawn, and produces a shining that is called phosphorescence. Lastly we have fluorescent light that is produced while minerals are being illuminated by ultraviolet rays. We all have at one time or another seen a rainbow. In this case the water acts as a prism and the photons of the light are separated. This produces the natural seven colors of our spectrum.

"It is commonly understood that light is always the result of the spending of some form of energy. Light from a star is produced by the reactions of the atoms and particles of gas of which the star is composed. In these ongoing reaction's, particles are constantly being transformed in to other kinds of particles and this change produces energy. Light from a star is one form of energy, produced by changes within the star. When these atoms and particles are exhausted and there is nothing left to make light, the star fades out."

Light appears always to die away, as in the case of the electric light bulb and the stars, when the source of the light ceases to create energy. This raises the question of whether the light from a distant star travels outwards forever. Outside our sun (RA) it is thought that light from the nearest of stars travels more than 25,000,000,000,000 miles before it becomes visible to our eyes. Does the light from these stars, out in deep space go on forever or does it "die" away? On this we are not sure, leading us back to the question at hand; What is the origin of the light?

To quote from an old, outdated *Encyclopedia Britannica* I am fond of using, "It might perhaps be expected that we begin by saying what light really is and then develop its characters from such a starting point. But this procedure is not possible, since light is essentially more primitive than any of the things in terms of which we might try to explain it. The nature of light is only describable by enumerating its properties and founding them on the simplest possible principles."

If this chapter is to make sense the reader must allow this text to become somewhat redundant. And the origin of the light shall be revealed. Let us repeat by saying that "light is the absence of darkness," the simplest explanation in the process of beginning to understand the origin of the light. This is the approach that was used in ancient Egypt. They believed in a universe of opposites; thus darkness becomes the "absence of the light." Although this is not a total explanation of the origin of light it is the starting point for what light is or what light is not.

Light is not the beginning of the universe. It is a mirrored by-product of that beginning. Within their religion is the belief that before the light came into being the darkness was a bounding. This is but one of many coded hints of the origin of light. If we can borrow from "existence" values that can in terms of the light, be comparable, this understanding becomes clearer.

To exist there must be molecular structure, as is found in rays, microwaves, space particles, and space dust. Some of these particles are subatomic and microatomic, without heat or visual movement. And there are atomic substances that have movement and heat and are electrically charged. And then there are those that are not either of these, but have an individual molecular makeup, each deriving from an original source, that allowed this development in stages. This means that as a rule, wherever you find heat and movement there will be light, or its mirrored image radiation.

Light comes in all colors, many more than appear in our earthly spectrum. Scientists today have a great understanding of white light. Thanks to scholars such as Newton and Huygens we understand that white light is actually seven colors enveloped as one. To see the colors we use a prism. The seven colors then magically appear as red, orange, yellow, green, blue, indigo, and violet. These are not the only colors dispensed by stars (light) within the universals pectrum. It is known that within the deepness of outerspace there are (exist) spectrums of light that range from black, faint blue, strong white, orange, red, dark purple, violet, blue, faint green, vivid yellow, and an abundance of others that remain unknown.

This would suggest that colors may be dependent on the the speed of light and that colors outside of our spectrum are traveling at a faster speed or slower speed than white [RA] light and obey different laws of refraction and diffraction. Or it raises the question, how can light be light and not be (white) light? We understand how light rays can be bent and influenced. The darkness of space has this effect, as does the atmosphere of the earth (a similar process isused in the creating of iridescent colors).

It is known that light travels in waves that can be grouped together by velocity. This is the reason that the spectrum of colors is revealed. When light or radiation is interrupted by the darkness of deep outerspace and particles are exposed to refraction and diffraction, it will no longer travel in a straight line. Rays and waves of light or radiation illuminate atomicand subatomic particles receptively making them visible to the naked eye and allowing this occurrence to be witnessed. The Aurora Borealis is a result of this natural phenomenon.

It was once thought that a vast distance between two points could not be spanned faster than the speed of light and that it is the fastest of all natural functions. The ancient Egyptians always believed that thought (the ability to think) was a natural function that occurs at speeds greater than that of light. If this is so, then it is possible within the realm of existence to allow natural and unnatural functions that are beyond the speed of light. Only contemplation of all the known attributes of light and rumors that maybe associated with light will provide us with the ability to determine the origin of the light. Nothing concerning its function or development can be rejected.

Alas we must relate to C.O.B.E. and its instruments that detected light and radiation. The three instruments it carried were the Far Infrared Absolute Spectrophotometer (F.I.R.A.S.), which made a precise measurement of the spectrum of the "cosmic microwave background radiation"; the Differential Microwave Radiometers (D.M.R.), which detected for the first time and was used to characterize faint fluctuations in the cosmic background corresponding to the density structure in the early universe; and the Diffuse Infrared Background Experiment (D.I.R.B.E.), which obtained data that can be used to seek the cosmic infrared background and study the structure of the Milky Way galaxy and interstellar and interplanetary dust.

Their measurements will greatly enhance this pursuit of the origin of the light. They were able to look back into the beginnings of the early universe, providing necessary clues concerning the origin of the light. F.I.R.A.S. and its spectrophotometer confirmed that the color red is indicative of the earliest light and radiation; that the heat emanated and collected was from a source further away than our own galaxy; and that red is the color of a developing universe. If there was a "Big Bang" than the color red is its signature, occurring before, during, or after this cosmic event that explains how it all began.

We have arrived at the very beginning of everything. Scientists that have analyzed this downloaded data all concur that this cosmic background radiation(Cosmic Egg) is the leftover of the Big Bang. In other words, this is where our search for the origin of the "Light" must begin. Unwittingly these scientist and scholars are applying New Age (Aquarius) results to "old age" concepts. Come what may, we have already embarked on a journey into the Age of Aquarius.

The "Age" of comprehension and earthly brotherhood (the end of racism), when mankind will once again come to know the real origin of the"Light." Recent findings may provide the missing clues, but not when bounded by the old (racist) rules. I highlight the old rules that have indicated that the ancient Egyptians or ancient Africans did not have a clue, or that they were incapable of understanding the true (doctrines of the light) origin of the light.

In the primitive minds of early mankind the light represented good and the darkness, evil. In these cultures, of the Nile valley, this predicated the need to choose between good and evil. This is the main difference between African and European civilizations, the way these two forces are perceived. If the light was good and darkness bad, then RA being the emitter of light and life, has already established itself atop the leader board, in the minds of ancient Africans millions and millions of years ago. After all, the life span of RA (Stars) is billions and billions of years.

The "Great Age of Pisces" began around the year 666 B C. Is it a coincidence, that civilization in Europe was just then emerging? European cultures are not cultures that embrace the sun [RA] and seem to lack understanding of the differences between good and evil. They have created their own deities that cater to their individual needs, respectively.

Most of them, such as the Greeks and the Romans, developed systems that are similar to the Egyptians' with different names that are a top the "leader" board. They have created their own (origin-less) cosmology.

They are the night cultures and have always shaded themselves from (avoided) the rays of the RA. They in their own way have replaced RA with fire and in the all the annals of human kind, are the last (continent) to accredit our solar furnace or seek the real understanding between light and all earth life (Renaissance, A.D. 1200). Their (combined) contribution to the modern world was a "ruthless" law that says, "Might is right."

According to the old school of European thought that is based mostly on the teachings of the Greeks, civilization and science and just about everything else began with the ancient Babylonians or a culture that was spawned in Asia (Mesopotamia). This concept is even pushed in modern-day America!

To challenge this popular concept, my radio carbon dating system will be cosmology and the origin of the light. I have mentioned the ancient Egyptians and their religious beliefs and will give reference to the ancient Babylonians at length in a chapter to come. But as I have studied hard on their views (Seven Tablets of Creation) on the origin of the light, in my mind they do not apply to this complete theme. Continuing this exploration of the past, let us now journey backwards in time before the Babylonians and into the minds of the ancient Africans who lived 2,000,000 years ago, and ponder the words of their chief deity:

"I existed in the primeval matter that evolved in the ocean of nothingness, eons of evolutions ago, before the beginning of time. When I thought the moment had come I fashioned my mouth and spoke my name as a word of power. With this power that was my birth right, I fashioned myself in form from the primeval matter of nothingness. I was alone at this time; there were no others with me then. I am the beginning of time, which began when I took form and commanded the power."

I intend to establish this ancient African religious document (Egyptian text) as the initial provider of clues needed in finally solving the mystery of the origin of the light. We will find by relating the concept of time to light that they also are comparable. After all, what is time? Time is light (the movement of heat and energy) traveling from point A to point B, or the measurement of energy (light) as it travels the vastness of space or in the case of our solar system, from the Sun [RA] to all (ten?) nine planets. Time is light, light is time, pure natural energy, possessing immense power, deriving from a source that was there in the beginning of the universe. Was this beginning of power in the very early universe the Big Bang?

THE POWER

"THE COSMIC EGG"

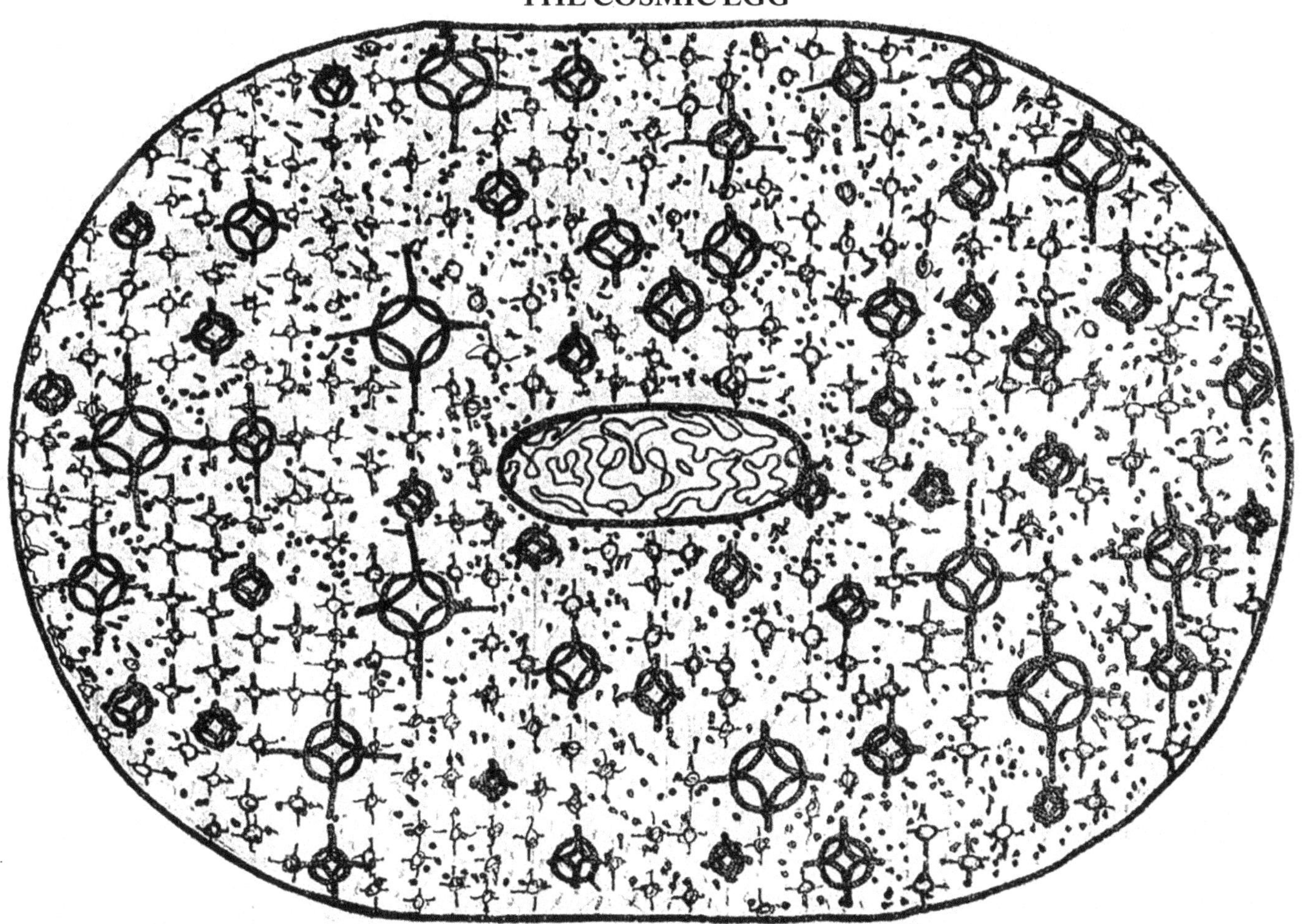

Surrounded by the ring of continuous gamma explosions.
(The reason that the Universe is still expanding!)

"The Earth Cannot Be Owned, Or Totally Controlled; It Can Only Be Shared."

Chapter 4
The Power

The universe is the source of all natural energy and power. It seems to have an endless supply of light and radiation. Before C.O.B.E., stars and the light from them were how the cosmos was perceived by the populace. It is a wonder that all can enjoy using one of the natural senses with which we are endowed, the gift of sight. We are born with this natural ability to look upwards at the night sky and observe the "wonder of it all." Before there was science or religion, observation of the universe was taking place.

As we have seen, not all light is white light. Would this suggest that different colors (spectrums) of light emitted by stars that are perhaps 25,000,000,000,000 miles away from the earth travel at speeds above or below the known speed of white light, producing these colors that are not, as of yet, found within white light? How much does the cold darkness of deep outer space concern this natural function?-273 cells

When we think of the universe, we think of power. This power is displayed in the emissions of stars and the light that they emit. And let us not forget radiation, the original source of heat and movement. How are these sources compatible?

The ability to produce heat and movement is one aspect they both share. If outer space is a void of cold, dark matter, than the heat of light will decrease as it journeys greater distances away from its original source. If the coldness of outer space has this effect of cooling off light, it should also cause a reduction or acceleration in the speed, raising the question: At what speed does light cease being light? The cold void of darkness is a drain on the movement of light. The emissions of radiation withstand the subzero temperatures of deep space far better. Of the two, radiation is the greater and it also can affect the speed of light. It is the force behind our expanding universe.

There is an interaction (constant) between the cold, dark void and the light and radiation that is effecting change. This is intrafarance, the ability of particle interaction, affecting the heat and the movement of light and radiation. It effects change. If we relate back to how the atomic reaction of particles produces energy within stars, intrafarance has this effect throughout the vastness of space and does not occur only within a star but also within the coldness of outer space, establishing the fact that pure energy can be influenced!

This may provide the answer as to whether speeds faster than light are-or are not-possible. Along with this increase of speed, time travel may also be possible. Change is a constant of the universe. This is the reason spectrums of unlimited colors are possible, because in this process of intrafarance, space is still creating itself (expansion). This is how and why it is still possible for there to be other colors within our own spectrum. It indicates that the process of creation is still occurring and that it is happening in a controlled, uniform manner. The ancient Egyptian creator of light says: "I placed all things in an order of existence so that together the stars illuminate outerspace and their movement is no drain on my [His] power."

The idea of the "Big Bang" becomes more questionable when the clues suggest otherwise. This theory that is based on the teachings of these same European and Western cultures does not give us a hint of the original source in which this "Big Bang" began. As its title would imply, it describes a universe that is expanding as the results of a violent explosion, in (total) disorder. Our universe is expanding very orderly.

It must be realized that to understand the ancient Egyptians, their science or religion as it equates to the "all, of alls" of creation (Hakka), it becomes necessary "to abandon the expectations, of finding within 'it,' or defining 'it' according to our own cultural, scientific or religious orientation." There is no modern terminology for this, unless it be God or the recent findings of C.O.B.E.-the cosmic background radiation.

The word *Power* as used by the engineer indicates energy under human control and its availability of being harnessed for the purpose of doing mechanical work. The ancient Egyptians as we are aware, were great engineers. The many surviving ancient temples and structures are a testament to this (Pharaonic Order).

They began the concept of measurement, starting with the tip of one's finger, progressing upwards to include the hand, the elbow, and the arm length. Perhaps in this way they contemplated the number of hands required to measure a distance as vast as from the earth to the moon.

I recall that in my youth, I was exposed to the "Hippie culture" (in the sixties) and how they responded when asked the question: What is the purpose of mankind? They would respond with answers such as "Wow, man! That's far out," and "That's deep, man, deep." Oddly enough, even today not many of us are aware of the purpose of mankind. The ancient Egyptians believed that mankind is the "measure" of the universe.

This identification is just another of the many clues, that imply an awareness of "outer space" and its properties. Of course, the knowledge, of the atom was known to them, ages ago, before; its so called discovery by the Greeks. ([see] Greeks and the atom) They sketched images of the entire make up of the atom. After all, how could it be a complete doctrine of "Light," without this knowledge? They developed charts that depicted, the likeness of "sub-atomic" and "atomic particles" and "anti-matter" and "radiation."

Could it be, they lived in past "Ages," of greater enlightenment? It is thought that Pharaonic Ancient Egypt began, in the great age of "Taurus." The era, of universal enlightenment however, began millions of years ago with the "Hominids" of Africa. These primitive inhabitants of the Nile valley have experience all the "ages" and more than once.

Cultures such as Greece and Rome and all other modem cultures are the result of the great age of "Pisces." Were the "Stars," in those very ancient skies more harmoniously aligned? The answer is "yes they were." As this age of "Pisces" fades out, mankind will no longer be plagued by "Its," celestial emanations of negative particles and negative thought(racism). If the destiny of humankind is to be the "measure of the universe," then humanity must reach an understanding that is above racial indifference. We possess very limited power, yet we are a special species within this universe, and we can effect change within it.

"SEB" 'THE GOD OF THE EARTH"

He is shown with the Planets life form displayed above hos outstreched arms and hands and also below encircling his feet. His body is replaced with the Earth and features "Africa" The very first continent.

"The Earth Has One Path For All; It Gives Us Life And Eventually Death."

Chapter 5
Seb

In all the known Universe there is one place that is most special to us all. That place is the earth or, as it was called by the ancient Egyptians, "SEB." This ancient word is the most generic of all words, when we are referring to this spaceship; everything about this vessel is special. It is a planet, of exactness, and to think that life began on it by accident is absurd, or that one race is favored above another is racially ignorant and totally and morally wrong.

Unfortunately, many times our actions are the results of faulty teachings that are wrong. This, "gospel" of dissension is preached within our schools and universities and seems to be co-signed by modern science and modern religions. These doctrines were good enough for my father. But are they good enough for me? Can the truth be compromised and still be truth?

In their religion that was fathered by science we have learned that Seb was one of the forms of existence in which matter appears. It began with Shu, the spirit of gas. Then there is Tefnut, his mate, known to appear in the form of liquid. Seb was a solid in appearance and the completion of the trinity of "Light." When we think of life, or the development of all earth forms, organic and inorganic, in the minds of many cultures (especially those deriving their origin in the great age of Pisces), one word is used to describe this magical process. God!

As we enter the new age of Aquarius idolic one-word definitions of this glorious development can no longer suffice. The idea that this unseen god created the earth and the heavens in the period of seven days is to me "far out" and lacking in depth. The actual understanding of the act of creation (which is still occurring) can only be perceived in steps that relate to an order that allowed the coming into being of the earth and the heavens and all the stars. The God theory of today does not address these issues in great depth if at all. Frankly it is to simple and does not come close to explaining the appearance of matter or the sources from which matter itself was derived. It is lacking in molecular structure.

The ancient Egyptian text states that Nut (the goddess of the sky) was the mate of Seb. And that this quartet, Shu, Tefnut, Seb, and Nut, represents the order of "one" becoming, three, then four, etc. That had to come to pass before the earth could successfully sustain life. In great detail, it begins the description of how this was accomplished. There must be three elements to complete the cycle of matter. Another is added ; air (Nut) or the atmosphere that surrounds and protects the earth from the harmful radiation of the sun [RA]. After these steps were completed, the planet was ready to support life and there was "fire," "air," "earth," and "water."

Most of the species of life that are derived from Seb and dependent on RA to sustain life, receive these gifts of life free of charge and are blessed by the numbers of days that count their (life) comings and goings, never questioning their origin. Only humankind has this need to seek his origin elsewhere. Modern science has created this need within him. Some would suggest that life traveled to this very "magical" planet from a source outside of our solar system and arrived here with the impact of a comet or asteroid. Is this possible?

Of course it is possible, but it's also so unpractical. For then the uniqueness of the relationships between earth and sun becomes lost and may allow one to imagine that all this happened by accident. Instead of the "order of existence" they create their own "disorder of existence." If the earth is the only place where humankind is found, it would seem appropriate that we protect it and cherish it and its life forms. It is our ignorance of the past that alienates modern man one from another, and now this ignorance would alienate mankind from his biological father and mother (the earth), Seb.

The "Big Bang" and other popular theories derived during "the age of Pisces" indulge in this alternate origin of mankind. This text demands more absolute, definitive answers than concepts such as "God" or the "Big Bang." It requires the readers to think for themselves and prepares ("new age") food for thought that goes beyond the boundaries of such childish, "old age," racist, "foolish-ness."

To continue: These four primary forces, earth, water, air, and fire, have been adjusting themselves, changing and re-arranging (accordingly) for millions of years before the appearance of mankind. Was all that has occurred over the last tens of millions of years merely changes of chance? If they occurred for the purpose of earth life forms (and they did) it should now suggest to the reader that there is an "order of existence."

There actually is an order of existence that is applicable to all things. "This order was issued to insure the survival of all things." "This is the 'Will' and the command of the 'Creator' of all things." Even their perception of (God) the 'Creator' was developed in steps. This insight that the Egyptians shared in perceiving, the concept of preparation before "existence," is lacking in other cosmologies. It sort of implies that the Egyptians were, in communication with the actual source and that the others weren't.

Is there a source that was not of this earth, but responsible for the formation of this planet? Yes! This source is the sun, our solar commander. The ancient Egyptian texts indicate that RA loved this special garden of life and endowed it with many of "his" own attributes, such as the ability to exist for millions of years and to continuously unbecome and then become anew. This is why the origin of the earth and earth life was a "done deal" within their minds. There is evidence of the connection between Earth and Sun everywhere. However, it is only to be seen by "the eyes that can see."

In the waning years of the age of Pisces, two new schools of thought developed. Each provided different answers to the question of origin. Science was one school and religion the other. These institutions found themselves in the position to influence the morality of humankind and from these schools of thought the Europeans developed ideas that have been passed down through the years. Even today they still hinder us (racism).

With a short review of the text, we will clearly see within our inner minds that the origin of the light is step two, in comprehending not only the establishment of the earth but the entire story of all creation. As has been said, this is a three-step process that begins on the earth. The origin of the earth is perhaps number three, in the greatest questions ever composed, and the mystery of the cosmic egg (cosmic background radiation) is number, one. Until the findings of C.O.B.E., the materials that fill the universe, light and radiation (that have emerged, as the signature of God) were related only in the ancient Egyptian text. These concepts were not evident in then-modern science or in the religions of today and are indicative of their limited, universal comprehension.

There are understandings that can be experienced culturally and passed on from one generation to another un-changingly. The "great ages" influence these changes that occur in the developments of earth cultures. These are the changes that we are examining, the progress and the reasoning that these cultures employed in separating the truth from the lie then, now, or at any time in the life cycle of humankind on this earth. After all, if we are the same species, why do we think so differently?

There are always institutions that managed to rise above other lesser schools of morality. In European cultures the Catholic Church provided the need, before breaking apart, fracturing, and forming other branches of the same concept, of monotheism. It was the birthing time of Christianity, and the Middle East and Europe are its mothers.

At this time the concept of radiation was unknown and indefinable to them, as would be this atomic era in which we now live. It would be like speaking of radio to Columbus back in A.D. 1492 and introducing a technology of understanding that comes from the depths of outer space. Would Columbus have been frightened of this new technology?

We remain (are) what we were, (non-rent-paying) residents of Seb. All earth life forms come under this umbrella, and try as you might, it will be (nearly) impossible to relocate this concept of existence, anywhere else. In all the millions of billions of stars in existence, only one special star features the earth, and before mankind, science, and God (as we understand him today), it was this way.

They (theologians) would have one believe that God alone was the one primary source behind the creating of the universe and the making of our solar system and mankind. Who is God? They defined God as a supreme "all-powerful living being" that is aware of all occurrences, including the thoughts of humankind.

What is his origin? He says: "I am (that) what I am." What is that? "I am the alpha and the omega." Or the "beginning and the end." Back in those days this was heavy stuff. This naturally prompted the question. Where was all this heavy stuff coming from? Turns out that it was the hand-me-down understanding that the Europeans borrowed from the Hebrews.

This deity (was able to) used his mind and power and thought the heavens and the world into existence, within a period of seven earth days. For them this explanation was enough and all the great Scriptures were written, passing on this eternal message. To challenge them or question these Scriptures was heresy. Punishable with death, for all minds open enough to desire more input or found in the pursuit of contrary information.

How could there possibly be another version of the truth that was unknown to this pious circle of Middle Eastern and European clergy? The reality of the matter is that European cultures have first-hand knowledge of "Jesus" but not of the one God of the Jews. As a result of this (?) understanding, even today, this is how God is visualized in the minds of believers. He is the alpha and the omega.

Modern science, which basically came into being as a result of the European "Renaissance," was having no part of this explanation that failed to answer the basic questions. In those early years, science was presented with more of the down-low approach, lest it too would fall under the guises of heresy. This sort of cramped true definition but eventually allowed for the growth of science. The scientists were allowed to discuss the theory of matter as long as God was credited. The church allowed ideas concerning, the changing, rearranging, and altering of atoms, but forbade this information to be written. In time it became possible to arrive at theories such as the "Big Bang" that provided "It" with the concept of molecular development and structure.

From these concepts of molecular development comes the theory that the earth is made up of debris that was leftover from the Big Bang. In time (by chance), gravity collapsed these molecules and particles into a nuclear core, encased by an outer shell, and in time this object grew into the earth. At this time, the earth was developing and so was our solar system. When this process was completed, an object from outer space (beyond our solar system) arrived and collided with the planet, creating the moon and seeding the earth with life. Now this is "far out"! Science has outdone religion with a theory that is just as "soulless" and empty. Both explanations are based on cultures that lacked a higher and "generic" understanding.

The Old Religion or O.R. taught the ancient Egyptians that our earth was once a part of the outer surface of the sun. So were the other planets that were captured by the gravitational pull of RA. It tells of a very violent period when stars and galaxies were forming and how the movement of RA's older and (larger) sister stars Alcyone and Sirius exerted such a great gravitational pull on our sun that energy was extracted from its surface and cast into outer space, where it cooled and eventually became our known solar system. Reason and accountability will begin to give definition and form to the origin of the universe if there is order in "Its" presentation.

It now should be clearer to the reader, that minds much more primitive than ours were able to attain a fuller understanding of this whole cosmic wonder and how and why this cosmic wonder occurred. Seb is the chapter that introduces earth life forms, among which is the "human animal." It contains a small fraction of "history," not just the story of man's development in Africa or Israel, Babylon, Greece, Rome, or Europe, for that is another subject. The "history" that is included in this text relates to science and religion and what the people of these cultures believed at certain periods (and "great ages") in their development of the cosmic understanding.

The Egyptian story continues: "After the planet had cooled and had become a stable environment, the earth was ready for life." It came in waves of light from our sun that seeded the planet and then it began to grow and flourish. The very first form of earth life was Flora (plants). Soon the planet was a living garden, almost completed and ready for mankind. It is at this time that evil, or the mention of it, comes into play. Let us not forget the concept of dualism, for it seems that as mighty as RA was, he had a mortal enemy.

This enemy was called "Apepi." To the ancient Egyptians, Apepi was the serpent of "darkness" that encircled the globe, preventing the establishment of life forms. When he became aware of the growth that was occurring below he was enraged and set his mind to alter or destroy what RA had issued into being.

They were two natural enemies, one a source of light and the other, darkness. Their battles are fought daily and the earth is the warzone. In their religion this explains why the planet rotates, because in their battles one was unable to vanquish the other. RA wanted to envelope the planet with the life-giving rays of light and Apepi wished to keep it in darkness.

Their battle was a stalemate and as a result of this, in order to provide light unceasingly to the earth, RA caused the planet to rotate, which began the cycle of day and night. It is said that because of this feud, the next form of earth life appeared. Apepi, introduced insects to the planet and their function was to consume all the flora that had come into being. The very first land battle began!

It was not within the heart of RA to destroy life and so he was hard pressed to counter this destructive new presence (insects) in a manner that would not destroy them. Thus he created a plant that would stop this new menace that was eating away at his garden.

The plant was hemp, a form of flora that had this effect: Once it was consumed by the insects, they became intoxicated and forgot their mission, and the plan for life on earth was saved. It was saved by the ingenuity of RA and this hemp plant that later would be referred to as the Tree of Life. The other life forms followed, each in order: fish, reptiles, birds, animals, and lastly, mankind.

This order of development is symbolically represented in art, in the form of the Egyptian ankh, or "key of life." All the species of the earth can be accounted for, by counting its six ends (points), culminating with the oval head that is symbolic of human life forms. If these ideas are given serious consideration it is not hard to see the merit within them. Slowly we are beginning to see how the secrets of existence were coded within these metaphors of religious doctrines.

The question of the origin of the earth or earth life cannot be answered by science alone or by religion alone. If we view earth life as being magical, as it is, then the religious approach is reasonable. Science becomes the molecular instrument of exposing its wrongs or rights. Today science and its findings are considered the real "gospel of fact." Complete theory only becomes fact by combining the two.

The phenomena of earth life are the results of reactions of living substances to radiant energy (photosynthesis) and depend solely on RA. the (Sun) So as we examine the text of the ancient Egyptians, it becomes absolutely clear that in primitive ancient Egypt, many times their religious concepts are expressions of scientific truths.

The late sir E.A Wallis Budge, keeper of Assyrians and Egyptian antiquities in the British museum, produced a book on the ancient Egyptians, that is commonly known as *The Book of The Dead*. From within this annunciation comes forth-with our text, that suggests a knowledge belonging to an indefinitely remote and primeval time. From this treasury of ancient thoughts rain down upon us the answers of the past, present, and future. The magical fairy tale of human life is one with the earth and we are all involved in this fable of stardust, with each individual holding equal shares.

The earth(SEB) and the sky (NUT) are both in their own way alive and teeming with life. Together they complement this ongoing cycle of reproduction. The earth is home to seven species that are indigenous to this planet. Humankind represents the highest form of earth life forms. The question I as know is to whom or what do we owe existence? Is it to a God, that was invented inside the minds of humans and has never revealed itself, or to the earth and the sun RA? If one can believe that life comes from the stars, as does all matter, then that question becomes moot.

In ancient Egypt this was a primary deduction and all their lives they professed unto it. With their acknowledgement of gods and goddesses, they became enlightened and by pondering thoughts (suggestive) of what they could and could not see the wonder of it all began to take shape and form, producing explanations that (examine and) explain earth lifeforms, "which are possibly the greatest accomplishment of it all." This was the purpose of the many elaborate and outstanding tales (gods) that revealed to the populace the story of creation. If we humans are made of star ingredients, then so is our Earth; It was conceived from a very special star RA.

The idea of a creator of the light and their gods and goddesses constitutes the beginning of science on this planet. Yet it does not constitute the complete beginning of religion on the planet. Both science and religion have their components that separate the two. And to define religion is much more complicated than defining science.

Religion today is so complicated and misconstrued that to correctly describe it as other cultures, past and present, perceive it, may be beyond my humble understanding. In this text I will try to explain what religion is and was to the ancient Egyptians and how it relates to the Cosmic Egg. In that sense a true religion must obtain its merit, by being established on fundamental truths (light and stars) that apply to all the creatures of this planet. In my mind, only a theory that combines science and religion can meet this requirement. It must be tested by the ages, and possess the finesse to endure for millions and millions of years.

It must have science, as in the story of the "creation." This understanding must be revealed. Mankind must believe this and then fashion a lifestyle (morality) that is adherent. It must include the art of burying the dead and thoughts of an after life, where only the worthy (those hearts that were scaled against a feather and in balance) would be granted life after death. It must provide doctrines that are a mandatory staple in our journey from the cradle to the grave. Finally it must answer all known questions of rebuttal graciously and be susceptible to the chaining of the Ages. This was the art of religion in ancient Egypt.

It was more suggestive of the art of science, dealing with elements that owe origin to the light, such as gas, liquid, and solid and even air. In this manner, they appear to have experienced science before developing a complete religious theorem. This was their way of alerting future generations to the relevance of science and religion. It would seem that we cannot have one without the other and expect order and enlightenment to be the result. If it is wrong then we must fix it. This new age demands that much from us, and it will accept nothing less.

Even a star with a cycle of "billions and billions of years" could die. It was written that: "When a star does not exist, any longer, then it has left its station, for 'He' (the creator) placed each and every star within the order of existence." Stars die, as does mankind, with death becoming the extension of the mysterious cycle of becoming and unbecoming; however, an afterlife is available for the worthy, beginning the cycle anew. This is a very complex understanding, which allowed them to formulate and develop the ideas of morality.

This is how and why science and religion were established millions of years ago in this valley that is nourished by a mystic river. In the Nile valley millions of years ago they buried their dead, instituting a ritual that separates the ancient Egyptians from other cultures of that period. It was their first step of many steps that mankind would make before becoming "civilized." They were the very first culture to call themselves "Men" and their mates "Women."

The ancient Bantu (a prehistoric African civilization) held the belief that the earlier beings were androgenous, man and woman in one person. They believed that a dance was performed around the "Tree of Life" and that after they had eaten of it (becoming intoxicated) and paraded around it several times they separated into two individual beings, one male, the other female. (Was this their way of acknowledging that there are three sexes?)

The earth today is battered and scarred, blasted and bombed and very nearly destroyed at the hands of (science) mankind. We threaten to destroy "It" and the sacred order of things that gives us life. "It" grows weary of religious doctrines that condone mass murder and threaten the innocent. This "odium" of religious theologicum threatens the destruction of us all, deriving justification from knowledge gathered in an age that forbade the truth. We are in the "Tuat" of ages, where the doctrines of might are justification of what is right. This separates us and allows the "One" to be against the "All." This darkness and madness will come to an end, and it will be soon and very soon.

Seb is the "special garden of RA" where men and women are "grown." So today as we ponder our aloneness, let us imagine the past as a door that opens to the present, providing us with information that ensures a more enlightened future. In concluding this chapter let the reader never again be amazed at the thoughts that were developed by these ancient worshipers of the light. They have provided us with a Grid (stars) on which to place their perceptions and their words, enabling us to separate science from religion, good from evil, and fact from fiction.

We can examine their beliefs that speak of matter and light together, and appreciate the science that was involved in formulating these ideas. Their concepts concerning good and evil, the way they envisioned the earth's origin and that of the light, cannot be taken lightly. We understand their concepts of how life developed on earth and was threatened, how this planet was saved by a plant, and how this plant played a part in separating men from women, and the fancy they employed in giving names to anything and everything. We can truly appreciate their symbolism and maybe, have a better understanding of them and their beliefs. We notice that always the concept of preparation is present. Flora is the original source of sustenance. Before other creatures could appear there had to be a food supply. Nothing happens by accident.

We can appreciate their practice of burying the dead and recognize that practice surviving today. Many of their concepts involving life after death are to be found in the religions of today. This afterlife was only for the worthy, meaning that it had to be earned. It pronounced the importance of morality and hints of the third sex. Their stories of the earth and its lifeforms continue and pass on and within this metaphor are large bits of the truth.

Lastly, let us receive the endowment they laboriously persevered for the generations to come.

"THE SOLAR BARGE" or the "BARK of BILLIONS and BILLIONS of YEARS"

Our SUN (RA) is shown seated in the body of a man with the head of a Falcon and the Circle and Dot, symbolic of the "Sun" surrounded by the Serpent "MAYHEN" He displays the Rasaurian Ankh, the Boat is Lotus shaped and the Bow displays the esoteric symbol of Billions of years. The boat soars across the "PERT" symbol of the SKY.

"RA Is The Emblem of The Creator That Brightens Our Sky."

Chapter 6
RA

Upon Seb, mankind and all other lifeforms have found a haven for survival. The provider of life then (ancient times) and now is RA. *The Book of the Dead* provides us with an abundance of information concerning this star RA. These ancient people felt a deep obligation to record all the information that was known concerning this solar deity. The tombs of the dead provide us with this story that was relished in those times, by the rich and the poor equally.

This is why it was so very important to associate life upon the earth with this star. It was a facet of understanding that was unknown (unrealized) by the European scholars who uncovered these findings. Only now with modern technology can we began to appreciate its importance.

Each new day was started with prayers to RA. Plate #1 begins: "Adoration's to RA when he riseth in horizons eastern of heaven." The word for word (Budge) retranslating sometimes is fuzzy, becoming necessary for this author to intervene and offer a simpler deciphering. The text commences with the Sun rising in the eastern sky. It speaks of RA being the vanquisher of darkness, the giver of life, he is great and he is immortal, living forever. He is the substance of all things (Star, Dust)giving to all freely and lovingly. He is a "god" that speaks (through rays of light) a god that can be seen and his warmth can be felt. This was their god, RA, from whence they and all things on the earth come. This is monotheism, with substance!

They had found in RA, a "Star" that is their commander of "Light" and life and all praise is due unto him. This was the beginning of the "art of prayer," and the ancient Egyptians were masters of it. The hymns to RA are so many in volume that I only mention a few within this book, those that pertain specifically to this text and those that are my favorites. The reader will see coded parts that pertain to the special-ness of the "Earth" and how RA set forth an order that gives it protection against life threatening objects from the depths of outer space; How the Creator of light is "self-created" within a black hole and how light and radiation are mirrored images of each other. From these ancient doctrines of light we will arrive at the answers to it "All."

The Age of Aquarius, mentioned throughout this book, signals a harmonic alignment of stars that include our sun. These stars open the doors that have for so long been locked against the truth. They trumpet the end of war, hunger, and man's indifference to man.

We cannot change the movement of the stars; we can only observe them. We can open our eyes and let the light in or we can keep them shut. The Age of Aquarius is symbolized by the figure of two men brothers or a man bearing a jug of water, and emphasizes his willingness to share. The passing age, Pisces, an age of darkness, was symbolized by two fish swimming in opposite directions (the sign of confusion). The great cultures that came into being under the sign of the Fish were dark civilizations, reflecting the darkness and the confusion that was prevalent in these times when the kings of the madness ruled.

Perhaps the greatest European observer was Galileo. Although he did not invent the telescope, he is credited with its discovery. In the year 1610, using his telescope, he found that there were dark spots on the sun and that these black spots moved across its surface from day to day. At this time, Galileo was employed by the church. He was a friend of and commissioned by the holy father, Pope Urban.

His observatory was located on the highest of the church's rooftops. There were precious times that he and the Pope shared observing the stars together. The people that held power at this time said what he called sun spots were due to faults in his telescope or in his eyes. They regarded his find as an insult, implying that the sun was not perfect, and this was heresy. If not for his friendship with the Pope, Galileo would have been tortured and after confessing to his error, would have been burned at the stake. This is what the truth was up against.

Pelades, a small constellation of seven stars, has a central sun called Alcyone which is orbited by those stars directly within this constellation. An observer from within the inner Milky Way looking outward would see an eight-star configuration of solar lights orbiting Alcyone. That eighth light is RA.

As our sun orbits around Alcyone, the earth passes through our Zodiac of stellar lights, taking roughly 2600 years to travel from one (great age) constellation to the other. The earth rotates with a movement called precession, which results in a slow westward shift of the equinoctial points along the plane of the ecliptic. This movement is similar to that of a spinning top as it loses speed, and it causes the polar axis to indicate a different point in space each day, requiring 2600 years to complete a cycle and return to any given position. If Galileo was aware of this movement of earth and stars, he could not reveal his findings.

RA has several movements. As does the earth, the sun also spins or rotates upon itself and in the same direction as the earth. Because of this movement we can notice a sunspot appear at one side of the sun, travel across its face, disappear for several days, and then reappear where it was first spotted. (Gallieo was correct on this.)

Besides this movement of rotation, the sun has a movement of translation, as it is called; in other words, the sun has an actual bodily movement, from place to place. All the other stars are also in motion. Our sun is moving through space at the rate of twelve miles (more or less) a second toward the constellation Cygnus, in the general direction of the brilliant star Vega. This point toward which the sun is traveling, carrying with it the entire solar system, is known as the Apex of the sun's way.

This is why the ancient Egyptians believed that man's destiny is written in the stars. Many of the elements known on earth are found within the gases of the sun (the stardust) of which we are made. It is believed that the sun is 81 percent hydrogen, 18 percent helium and 1 percent oxygen, magnesium, nitrogen, and others. But you must not think of these elements as they are found on earth. Even those we know of as solids exist there as gases. Iron, for instance, is a solid and (when heated) liquid in normal earthly molecular structure, but within the interior of the sun it is a gas- once again suggesting the complete dualism of sun and earth and that one was once apart of the other.

RA is so huge that save jupiter our entire solar system would fit into it several thousands of times. However, It's small, in comparison to other stars. Its rays of yellow and white light are unique, providing that special energy that allows the earth to retain life and to remake itself for millions and millions of years. We travel through space on Seb, the most reliable of "spaceships," and we go where RA brings us.

In establishing a religion, the understanding of the stars is a necessity. That comprehension begins with RA. In this regard one must admit that RA is our true commander. This knowledge of earth and sun and that of other stars and constellations of stars was banned and purposely destroyed. It was first misunderstood by the Greeks and then rejected by the world-conquering Romans, who showed a special disdain for this ancient African culture. In my mind this is the beginning of racism, which later would become a very common European practice.

Thousands of years ago, before the invention of the telescope, humankind had been observing the stars. There were two great civilizations that were noted as observers of the night skies, one African and the other Asian. It was known that both collected knowledge of these observations. In the fashion of the Greeks, the Romans also chose to adopt the views of the Asians. This was a dark decision. Their might made it right and this conclusion burdens us even to this day.

Today we are afforded the luxury of reexamining their theories and results, highlighting those events in the history of gods and men and of war and of conquest that would shape the future. Many European and American scholars believe that the ancient Babylonians were the most correct of the ancient sky observers. Some relate the date 2100 B.C. and the ziggurats of Ur as the earliest observatory of the stars.

The Bible relates that it was a time when the Jews were being held hostage in this ancient culture that thrived between the two rivers called "Tigris" and "Euphrates." This was an era when gods clashed, when mankind began to witness the end of polytheism and the old cadre of gods that had existed for millions of years, giving way to a new and less complicated idea that expressed the concept of "All" as only "One." The theory of monotheism began to announce itself. The theories of one god were at war with the theories of many gods. The Age of Pisces, also known as the House of the Jews, had begun.

Just saying or believing it is so does not make it so, or does it? It was a very confusing period. Whom could you believe? Who was telling the truth? In times such as these, the dark times that opened the Age of Pisces, even the gods were at war with one another and ancient Egypt, Israel, and Babylon became the sites of the battle being waged. It was the God of the Jews against the gods of polytheism, the Egyptians and the Babylonians.

The Babylonian king at that time, Nebuchadnezzar, was a "Great Warrior." He and his troops were returning home after ravishing Egypt. Along the way he subdued the nation of Israel also. Taking many Jewish prisoners, he vowed that nothing short of an act of God would obtain the release of his captives, Three hundred of which were taken out of Egypt. Monotheism, a theory that is completely without science, begins to replace polytheism. The millions of years of generic understanding were about to be swept away.

To the Babylonians the mud-bricked stepped towers or ziggurats were symbolic of the power of mankind and were their connection to the gods. They served not only as places of observation, but also as temples or the place where their great god Marduk would dwell. These structures were representations of the awesome power and knowledge that they possessed.

Their cuneiform script contains over five hundred wedge-shaped markings or scratchings. When you consider the Hanging Gardens and their knowledge of science, philosophy, religion, botany, and just about everything else, this was a remarkable Asian culture. Ancient Babylon was said to be "a beautiful city of the light that shines like a pearl in the desert" and the enlightened ones there were said to be many.

The results of these observations produced a large volume of plates written in the ancient cuneiform script. They are credited with accurately predicting the cycle of the moon eclipse that occurs on this planet, concerning the movement of the earth and the moon as they revolve around the sun and each other. Be what it may, polytheism did represent a knowledge of science and a broader understanding than its executioner monotheism.

To the ancient Egyptians, the circle and the dot in the sky represented the signature of RA. And for millions of years, these eclipses occurred in skies that were above the Nile valley. Nearly 3,000 years ago another Asian culture, the Chaldeans, had already discovered that eclipses recur at intervals of eighteen years and between ten and eleven days. The knowledge of the stars was much more observable in polytheism. This observance is not to be found in monotheism.

The Babylonian script speaks of a hero god named Marduk that was a creator of heaven and earth. He was their god that ruled over the earth and its moon. His attributes were similar to those of the Egyptian creator of the light. In the divine order (the concept) of the gods, even in Babylon, it was known that Egypt was the home of the primeval God. So many of these Asian's views were based on knowledge that was of African origin.

Evan Hadingham writes in his book *Early Man and the Cosmos* that over twenty-five hundred years ago immense manmade towers of mud brick rose up toward the night skies of Mesopotamia. Here in the "land between the rivers" of present-day Iraq, the towers or ziggurats must have provided perfect observation points for the scribes, who had developed an unrivaled knowledge of the sky.

Hadingham goes on to say that "the great king Nabopolassar, father of Nebuchadnezzar, ruled from 625 through 605 B.C. In an inscription he tells how in a dream the hero creator Marduk instructed him to (restore) the tower of Babylon, which had become weakened over time and had fallen into disrepair; he commanded me to ground its base securely on the breast of the underworld while its pinnacles should reach upwards to the skies." This great structure connected the Babylonians to the divine order of the heavens and came to be considered the "structure of great arrogance," or the Tower of Babel. It and Polytheism would share the same faith.

Hadingham writes, "The ideas of the Babylonian astronomers are particularly important because their highly precise predictions lie at the root of Western science." So, many of our modern-day writers are compelled to pass on this hypocrisy. It seems ridiculous to me that back then the Greeks found more merit in the findings of the Babylonians and not with the older findings of the Egyptians. I suspect that it was a logical way of appeasing the Romans who had also conquered them. The knowledge that Plato carried from Egypt to his countrymen in Greece is not reflected in this decision. Instead it was how the Romans also accepted these findings and how they became interwoven into our modern world.

Babylon, Greece and Rome, each nations of conquest, all made in the image of one another, were takers and often claimed credit for the ingenious developments of others. But since they did not understand in full, their undertakings were partial and not truly complete (or applicable) in relating backwards to the original source of and reason for development. These were the cultures of logic. One paved the way for the other. Their so-called wisdom was not indigenous to their own undertakings (lacking the experience of trial and failure) but adopted or stolen from those that they managed to conquer. Was this not the European way?

This explains why these non-generic warrior nations adopted a calendar that was based on the cycles of the moon. And as it turns out, these cultures never had much success with this system. The reason for its shortcomings is because the cycles of the moon are not constant. They vary between 29 and 30 days from new moon to full moon. It seems that knowledge of the eclipses and moon cycles are not fundamental enough to establish a seasonal calendar. Eventually they came to understand that a true, accurate seasonal calendar could not be based on cycles of the moon.

Finally there was something that could not be resolved by arms, might, or war. An accurate calendar, then and now, allows man to keep pace with the seasons. This requires knowledge of the stars and an understanding of them. And in ancient Babylon this was not their main strength. Their universal observations were insufficient concerning the knowledge of the stars. It rather seems to begin with and end with the observations of the moon. They were unable to produce a calendar based on their star observation, not understanding fully that time must be equated to the light of stars that are more constant than the cycles of the moon. Daytime and nighttime are the results of the earth's rotation as it revolves around the sun, which is a star.

It was not until 44 B.C. when Julius Caesar and his armies conquered Egypt that this concept of measuring the length of seasons by calendar was partially understood by the Europeans. After that conquest he adopted the ancient Egyptian solar calendar that was correct to the hour. It was based on the appearance of and disappearance of stars. (See the Rasurian solar calendar.) The mighty Caesar altered this ten-month Egyptian calendar of 365.25 days and added to it his own month, now called "July" (for Julius). His god son Augustus Caesar added his own month, "August," we now have the modern-day twelve-month calendar, adopted by the Romans and fashioned in the minds of ancient Africans.

There are and have been other cultures, such as the Chinese, Inca, Mayan, Jews, and Muslims, to name a few, that continue to use a calendar system that is more indigenous to themselves. But thanks to this Roman conqueror the world in general is unwittingly paying homage to the "Dark Continent" in its observance of this worldly accepted twelve-month calendar.

As we have seen, when conquerors adopt systems that are not indigenous to themselves, usually a gap in understanding appears. Such is the case in our own modern-day calendar. This gap is known as Leap Year. This promoted an in exactness of measuring time and staying in order with the seasons; centuries later the Catholic Church was required to intervene.

In 1582 this "gap" revealed itself. The seasons did not match with the calendar. Pope Gregory XIII was forced to reform it. he and his scholars adjusted it so that it would not need tending again for 6,000 years. And we now have the Georgian (corrected and updated)version that still contains that element of in exactness, Leap Year. A legacy that was for the whole of the earth was interrupted and misconstrued to apply to the needs and understanding of the European conquerors. If it works, why fix it?

The knowledge that the ancient Egyptians inserted into their calendar was an understanding of the stars. Even the great Julius Caesar was intimidated by this wealth of cosmic understanding and upon his order the great library of Alexandria was burnt. Within that library were ancient scrolls that contained the bulk of knowledge concerning the whole of the universe and on every special book already noted, *The Book of Great Enlightenment.*

The practice of destroying knowledge by the burning of books begins here with the mighty Caesar. Many centuries later another conquering European nation, Germany, and its Caesar, Hitler, would inflict on the Jewish race the pains of captivity much the same as was as done to them by the Babylonians.

He also would be a Caesar and have his "Sauberung" night of burning books in hopes of purging the earth of Jewish in sight. In his vision of a white Aryan Nation that was to last for a thousand years, and was to be free of Semitic influence, Adolph Hitler was a racist. He was a hero to some and a madman to others. It is true that when we are ignorant of history, it will repeat itself.

It does not seem strange to me that the downfall of (these) nations past, present, and in the future will most likely continue to be repositions of the past. In today's world we are living the great Lie that was perpetuated in part and in full by the nations that have been mentioned, the results of which produced racism. In the eyes of RA we are all equal. "A house that is built on sand will not stand forever!"

The Romans copied many things from the ancient Egyptians and while they were superior in the way of war they were in the dark when it came to understanding the doctrines of the light. Their chief deity was Jupiter, with a mere planet named for him. He was chosen above RA. This ignorance is reflected in all cultures that owe origin to the Age of Pisces.

How would one explain to any of them what I am attempting to explain to the reader? When the telescope was used to explore the heavens its findings were considered heresy and these findings, although correct, were denounced by the Church.

Generically, a real church is a mutual collection of ideas and thoughts that are applicable to the understanding and the benefit of all humankind. Its doctrines must not favor one above the other but spread the gospel of equality. The strengths of a real church are derived from its truths, drawn from knowledge of the stars and from their gospel of billions and billions of years. This was not the position of the Catholic Church (or modem science). Anything that did not fit into the limited minds of those who held power was condemned and rejected.

Out of Europe came these messengers of the darkness, those that would remake the world by might! This was the first European "Inquisition." They would enlighten the world with knowledge of Caesar: Accept him as your deity or die. His "Jesuits" were carried on the sides of his soldiers. Their gospel was conquest. Their ignorance of the light, or the absence of the light within them, allowed them to rape and pillage the earth and subjugate humankind. There was no pressing urge for science as it relates to the stars. Their science was developed around the pursuit of war, producing efficient fighting soldiers that were without morality or brotherly consciousness.

Rome was the center of prejudice, especially against Jews and Africans. When Caesar married the famous Cleopatra, she bore him a son. This great African queen and her offspring were rejected and Caesar's involvement with this woman was cause for him to be murdered. It was said that Brutus, who loved Caesar, joined the conspirators rather than accept Queen Cleopatra or have a half-breed possibly succeed Caesar.

The pursuit of mankind, with the intention to murder and kill, simply because of color or race or religious beliefs, begins in Rome and with those that would murder a Caesar. Racism began in depth in ancient Rome. The Romans' passed on their shallow views on morality to their European neighbors and it was passed down by them to us today.

In concluding this chapter, let us realize that Rome or the continent of Europe itself was a place where knowledge of the light and stars was rejected.

The plains of Existence are more numerous than our dreams. In order to exist within the ethereal plain there must be heat and movement. Heat and movement are found within the photons (particles of light) and in the dark black matter of radiation. From these two, we will seek the answers to everything! The hidden message that was recorded in the *Book of the Dead* once again becomes an avenue of enlightenment.

Many thousands of years ago, the high priest asked this question of RA: "From where do you come?" RA answered; "I am self-created and the emblem [the disk] in your sky that is symbolic of the creator of the light." RA is a story that leads to the Creator that birthed the light and created the universe of light and, as we shall see, existed within a Black Hole. Ausaures, their first high priest, was such a man. He was primed from birth to bring the knowledge of the stars to his brothers and sisters.

We will now go into depth in explaining this mysterious Creator that reveals himself through knowledge of the stars. "I am he that existed eons and eons of evolutions ago, before the beginning of time." The idea of existence before time (light) hints of the encoded message that reveals a cycle of perpetual life. If time is valued in terms of light, then this doctrine speaks of lifeforms that existed before time itself (before light existed) in avoid of subatomic particles. This is the domain of inner space, already existing before the emergence of light. This would suggest that there is an inner and an outer universe. One allows the existence of subatomic particles and the other of atomic particles. Both exist apart from each other.

Today, because of the microscope, we are aware of the smallest of cosmic particles. Before Dr. Louis Pasteur awakened within the minds of the Europeans the concept of unseen microscopic bacteria that carry dreadful germs (and at first they thought him mad), this concept was completely unknown outside of the ancient schools of light. Plato noted that all the priesthood of the Egyptians had clean-shaven (heads) and how they thought that the body's hair carried unseen germs. Bathing was a national pastime in this desert land of the Nile. And that the Greeks and Romans adopted this practice of bathing is evident. Whether or not they were conscious of the unseen bacteria, their everyday practices do not support this idea.

Possibly, the very first Europeans to show an interest in atomic structure were the Greeks. R.A. Schwaller De Lubicz writes, "After Zeno demonstrated the nonexistence of multiplicity in order to proclaim a unity devoid of all abstract and theological character, there remained for Leucippos but to prove logically that the whole is composed of an infinity of particles." He goes on to say, "It is impossible to separate Leucippos from Democritos; the latter is traditionally acknowledged as the father of atomism. None of their writings have survived, and we must turn to Aristotle in *De generatioe et corruptione*, another text that comes down to us only in Latin, in order to learn the principle of this first doctrine of the atom."

So it should become evident to the reader that if we must rely on the Greeks or Romans for an interpretation, we can expect gaps. They were unable to understand in full, and the understanding that they pass down is partial. To continue, according to De Lubic:

> Here is how Aristotle summarizes the Leucippos-Democritos doctrine: Leucippos (himself coming from the Eleatic school and contrary to the Eleatics who proclaim Oneness) believed he had sufficient reason to establish a theory in accordance with the facts supported by sense-perception which would accord with production and destruction, motion and the multiplicity of things.

> This much conceded to phenomena, he granted to the Monists that motion is impossible without the void and that the void is a non-being, and in no way participates in being.

> According to him, being, in the strict sense of the term, is an absolute plenum; as such, it cannot be "one," but is composed of an infinite number of elements which are invisible owing to their smallness.

> These elements move in the void (for the void exists); by their coming together they cause coming into being, and by their separation they produce corruption.

They act upon each other according to their mutual contact, because the manner of this contact is not uniform; and, in combining and intertwining, they generate the universe.

Indeed, from genuine there could never emerge a multiplicity, nor unity from multiplicity properly so called: That is impossible.

De Lubicz says:

From this text it is conclude[d] that the atomist[s] take into account the evidence of the senses and that they acknowledge multiplicity and motion; motion in turn, presupposes the existence of a void. This theory recognizes the unity of being and the multiplicity of things.

Nothing can be created out of nothing, nor can it be destroyed and returned to nothing. All change is but the aggregation or the desegregation of parts. Nothing exists but atoms and the void: All the rest is hypothesis.

The atoms are infinite in number, and their shapes are of infinite variety. The differences in all things stem from differences in number, size, form and arrangement of their atoms. The atoms have no qualitative differences: they have no internal states: They act on each other only through collision and blows.

De Lubicz concludes: "The Greeks are wrong to recognize coming-into-being and perishing, for nothing comes into being or perishes, but rather mingles or separates from things that are. Thus they would be right to say composing rather than becoming and decomposing rather than perishing." A letter from Ascdepius to King Ammon opens with the formal recommendation that an important writing not be translated into Greek.

It would seem that even the Greeks and their logic were unable to unravel the mystery of the atom. The existence of unseen bacteria and invisible germs that can do the body harm was unknown to Hippocrates, and he is their father of medicine. There are times when a culture is given too much credit, while others that are more deserving are discredited. The great age of Pisces was one of those times.

We have seen how very few documents written by Greek Atomists have survived, and how Aristotle had to retranslate into Greek these concepts that were written in Latin. The theories are quite logical even if they are not always applicable or correct. It seems that even when they were not sure of what they were speaking about, they were able to do so logically.

Now, can the reader imagine explaining the principles of light and matter and of unseen atomic particles that create bacteria to a council of the holy fathers back in the time of Galileo? When it was a fancy that the earth was flat and that the sun revolved around the earth. Unimaginable-that's what it would have been like in that Age of Pisces where you would have encountered a predetermined justice that was based on might, not right. One answer must now be accepted as an apparent truth to us all, and that is that without RA we would not exist.

Their creator of the light claims to have existed within the primeval ocean of darkness, a place where subatomic particles and atomic particles converge, where in matter can be created or destroyed. (The blackhole is such a place.) He says: "I existed in the primeval ocean of nothingness." Many are the years that I have pondered the existence of such an ocean. For I felt that this was the validation of this story that involves the everything within the universe.

Thanks to the power of the supertelescopes we now are able to observer these massive "black holes" that are quite abundant within our universe. In the center of all the known galaxies you'll find a black hole that is one half of one percent of all the mass within these pools of light. Under normal circumstances matter cannot be destroyed, only changed, rearranged, or altered. These massive oceans of dark black antimatter possess, an uncanny power unheard of until now (modern times) that has the gravity and the resolution to absorb light and dispense the light.

This is the scientific explanation of this most ancient of Egyptian texts, which is actually telling us of the secrets of the universe. This is the proven scientific process of how stars are made and should answer once and for all, the question as to the origin of the light."

"THE CUBE OF EXISTENCE"

It displays the concept of Light, Black Holes, and the C.B.R. or Cosmic Egg,
The door of Thought is partially displayed completing the Cube of known
Universal Phenomena

"RA Is The Star That Announces The Dawning of All The Ages."

Chapter 7
Existence

If we people did not exist there would be no need for science or religion. We are the spirits of heat and movement and without us, there is no reason for the pursuit of knowledge. Some people (cultures) are enlightened by knowledge and others are not. In ancient Egypt they developed (adapted) their culture in accordance with the "Great Ages" and the constant belief in an order of existence. There are three forms of Existence. One is light and two is Radiation; the third is the Spirit. We are the by-product of these cosmic wonders. The spirit of existence is "Intrafarance." It's understanding requires a layer-after-layer cocking effect.

Science and religion are the results of people and their understanding of knowledge. The understanding of the ancient Egyptians, concerning the unseen, subatomic particles and atomic particles opens to us a door that allows us once again to pass through, to the other side of "Existence." These particles are (emitted) from the source of everything or the "All of All's." They are from the source that "existed" before the "Light."

Let the reader not be, frightened as we pass through this door that, demands the ultimate in our usage of brain power. This expansion of thought allows a surge of generic understanding, that was experienced by these ancient Africans and can now be reasoned, explained and experienced by all. They believed in the power of their creator of light as being a concept that is fully comprehensible. They also believed in a power that was not fully comprehensible. (Radiation) There is a text that reads: "All life is derived from power, incomprehensible. So is the universe and all the powers, involved in creating this universe."

In this chapter and the chapters to come, we will try and expand our thought level and begin to explain that which cannot be easily understood. The creation, theme, explains how the laws that govern the light are bound to yet another set of laws. It is within these (mirrored) laws that we will find the answer to all of "existence."

The will of power that is comprehensible is the power of the stars. The will of power incomprehensible is power complete. This complete power is called the Primeval Germ. We shall relate to it as the Anomaly that (the spacecraft) C.O.B.E. rediscovered, the cosmic background radiation. The rediscovery that modern-day scholars, scientists, and researchers agree is the very beginning of the early universe. The beginning of Existence.

These early Africans seemed to have a working model of the entire structure of the universe, something that even today science and religion are missing. In this book we attempt to recreate that model in thought and in art form, specifically developed for this text. (It is thought that the universe has no center of beginning)

Both science and religion agree that the beginning of the universe required enormous power. This power was radiation, the mirrored image of light. Their creator of the Light provided the clues to understanding the part played by the blackhole so that we actually comprehend that light can be destroyed and light can be created. Also their belief in a primeval germ accounts for the origin of radiation, leading us to eventually realize that from its emissions and evolutions comes Intrafarance, the reason for the depths of space itself and the story of everything that is or ever has existed.

The black holes are the doors that open to oblivion, in which the light enters but never leaves. If this is so (Hawkins radiation), then it is the resting place of the light when it dies. In order to have death, there must first be life and a cycle that perpetuates both, Intrafarance, indicating to me that there could not have been a "Big Bang" that started it all. There are no random happenings within the universe. Everything must relate to an order of existence. This concept is fact!

The deductions involved in this Big Bang theory indicate that the universe is moving apart uncontrollably, and that because of the laws of gravity the expanding universe must reach in its own way or experience a "Helio Pause" (as with energy emitted from RA). And that upon reaching this point it must collapse once again on itself. This concept involves energy that is violently discharged outwardly (and) then later will collapse on itself and the Big Bang begins again. It seems to hint of the idea of beginnings and endings and perpetual existence.

In this way even the Big Bang theory is reflective of the laws of dualism. C.O.B.E. has alerted us to the fact that ninety percent of the energy in the universe was released in its infancy. This coincides with the Big Bang theory and although it has its gaps, modern day scientists reluctantly support it.

Until recently it was thought that maybe the universe is not still expanding, but receding. After decades of pondering this dilemma it was established that the universe is still developing harmoniously. The Big Bang mentions how the universe can begin and gives us a good explanation and estimation of its ending. However, it does not present the ideas of origin.

It seems that stars and galaxies of stars are being swallowed by black holes. Their life cycle of billions of years is threatened by a force not explained in this theory. It does not speak of an order of existence nor does it explain its own origin. The Big Bang is not a theory of self-creation. It is a theory, like so many other Piscean Age theories, that contain too many gaps. The universe in which we live is the result of the most orderly of evolutionary beginnings and endings, and the only gap is our level of comprehension.

Today it is known that the universe is still expanding and in uniform fashion. The Big Bang theory does not explain itself or, as we can see, where the universe is going or from whence it came. Back in its day it was far out. It was "deep!" -but not that "deep!" There are more things in reality that are already "existing" than most of us will ever understand. And there are new things that will come into existence. These things can be perceived, but never fully understood, Leaving only the concept of self-creation as plausible, if not always comprehensible: *Intrafarance*, the process that allows the universe to continuously create itself.

This text introduces a new concept of how the universe is continuously remaking itself. The word that describe this comic wonder is "Intrafarance" (The Spirit) and as we shall see it is applicable and vital in understanding, the beginning of the universe and how "It" still is in the process of universal development. "Hakka" is another word that may be unfamiliar to the reader. It is the complete concept of how it all begins. It starts in "Oblivion and ends in Oblivion." This chapter prepares us for what is to come. The "existence" of life forms on both sides of "Oblivion."

Faith, Forgiveness, And Redemption Are Colorless, But They Are Not Invisible.

Chapter 8
C.O.B.E.

The Cosmic Observatory Background Explorer and its instruments measured the brightness, the temperature, and the spectrum of light that was emitted by this "cosmic wonder," (CBR). The diffuse infrared background experiment mapped the absolute sky brightness in 10-wave-lengths bands ranging from 1.25 microns to 240 microns. This data contained the signals from the cosmic infrared background and the foreground emissions from extra galactic sources, our galaxy, and dust and other sources in our own solar system. The object that was being observed was not a star, yet laws that apply to stars and light also applied to this great source of radiant energy.

The differential microwave radiometer measured the radiation that was distributed when the universe was still very young. This source of cosmic radiation found within the innermost part of the universe, is the natural source of everything. It is the primeval matter of itself and is self-created. It derives origin from the all-encompassing void, known as oblivion. This wonder of wonders is emitting rays of radiation that are older than the light and, as we shall come to understand, is the reason why the universe is expanding.

The Far Infrared Absolute Spectrophotometer measured the spectrum of heat and energy that was released within the first year after the Big Bang or the energy released in its evolution of itself after it escaped oblivion and took form. Finally there was proof of the beginning of the early universe. In the minds of most scientists they, had found God. He could be seen and felt and even explained, giving them the opportunity to find out what it is that makes God God. (As of this writing, not a single soul has discovered this.). If we are to attempt to understand this most natural of happenings, intrafarance, than the interpretations of the ancient Egyptian text, as I have reinterpreted it must now be applied.

Let us say that light is the result of atomic activity. When the electrons, protons, and neutrons are stable around the nucleus of the atom, light is produced. If the electron leaves its orbit and is drawn inward towards the nucleus, the atom will radiate, and when the electron is pushed away from the nucleus the atom absorbs. In order to understand this cosmic wonder (radiation) we will first find its likeness in the light. And nothing can be dismissed or rejected until it is found to be not relative to this source of cosmic radiation.

If we apply these laws of the atom to other cosmic structures, the mystery begins to make sense. For instance, light dies; within a black hole it is absorbed. Does this not suggest that within a black hole the electrons are responding in accordance? Within these blackholes, matter is being created and destroyed. This identifies the electron as the "staple" molecule of both light and radiation. When these electrons inside the blackhole are completely spent, even the blackhole fades or is absorbed into oblivion.

If this anomaly-the cosmic background radiation or Cosmic Egg-existed before the light before the Big Bang and it still is existing, then without a doubt it approaches the value of and the title of "God." It causes me to ponder strange thoughts that become even stranger questions. Would the God of the Jews have been able to explain this anomaly to Moses? Could the mighty Caesar or the nefarious Adolph Hitler have been mentally capable of understanding the findings of CO.B.E.?

It was a spacecraft that revealed this revelation to us? Who informed the ancient Egyptians? Clearly the Doctrines of the Light illustrated in the *Book of the Dead* are right on in describing what it is that the spacecraft located. This information was recorded within the tombs of long-ago dead African kings. The walls of these tombs were inscribed with the sacred words of life and death and the story of stars, explaining why they aspired to rest in a place that was housed in Eternity, an underworld that was ruled by Ausaures. Was this that special place? Strange, is it not, that knowledge of the stars was buried in the bosom of the earth and in the house of the dead?

Let us now digest mentally all the pertaining information that was downloaded to N.A.S.A. According to the Microwave Background Anisotropy, fluctuations are a natural outcome of the physical processes in the early Universe Starting with super large scale "fluctuations" in energy density or entropy, gravity and pressure forces, lead to acoustic oscillation in the photo-baryon plasma. ("Baryon" refers to one of the ordinary atoms in the universe, they are mostly hydrogen and helium.)

Until about 200,000 years after the "Big Bang" the hydrogen within the universe was fully ionized (electrically charged) resulting in a high, Thomson opacity (in other words the plasma was opaque, like the interior of the sun.) About 2,000,000 years after the big bang, when the temperature decreased to about 3,000 kelvn (very hot) the electrons and protons, combined to make neutral hydrogen, a colorless transparent gas. Looking outward in space within the microwave part of the electromagnetic spectrum, we look back in "time" to when this recombination took place. Science has finally discovered the origin of the "Atom."

These are the findings of C.O.B.E. as they are related to the Big Bang. They describe how the atom developed and how that was a reaction of the fluctuations or emanations of this primeval source of radiation. If the Pope were to be informed of these findings and then asked the question of God's origin, how would he answer? "I am what I am," etc., will surely not suffice. We are entering the age of enlightenment and I pray that he also will be influenced and enlightened.

Today the scientist would explain how the structures seen in the universe were formed, due to "self-gravity." Initial small perturbations in matter and radiation were gravitationally amplified into galaxies and clusters of galaxies, etc. Religions of today have an old enemy with modern instruments (super orbiting telescopes) much more powerful than the one Galileo and his friend Pope Urban enjoyed.

The popes of yore were reluctant to accept Galileo's finding that the earth revolves around the sun. I wonder what they would do and how they would accept this information provided by C.O.B.E.? This leads me to ponder the question: Will history repeat itself?

It would seem that in today's world, science has developed beyond religion. It now attempts to define God! Yet it still has not provided all the answers or a working model of the universe. With the secrets of the early universe before them, they continue to be puzzled. They have found the all of alls but cannot truly comprehend that which they can perceive. They are very much engaged in developing a sound theory of how it all works.

Their search for the beginning of the universe (of light) starts with the hydrogen atom. Could this be the reason that they have so far failed? We are now observing the cold, dark matter of the early universe. The cold and mixed dark matter and the hot matter and all the possible combinations of matter are revealed to us, the gifts of the super telescopes. The scientists of today are exposed to the very essence of God. They aspire to establish a communications link with this living (egg-shaped) source of radiation.

Many throughout history have contended that science was designed to remove man from the essence of God, while in reality it has drawn us closer. It can no longer be said, by a (thinking) individual, "I don't believe in God." In the past this could be excused. Why believe in a God that is hidden from us and always unseen? And although you disagreed with this opinion, it could out of fairness be respected. The end of the era of atheism and nonbelievers has arrived, along with the great age of enlightenment!

Science has proven that God is observable. The "he" or the "she" or the "it" was always there, unseen by the naked eye. That primeval instinct first observed in the lower forms of earth life, acknowledging the wonder of God, is an ancient truth that now is a fact. When science questions this anomaly, it receives answers that derive from the birthing cradle of the universe. Now, because of C.O.B.E. science is prepared to answer most questions that can be formulated.

It would appear that the question of matter and its origin have been answered. These answers cannot be rebuked by religion that has stood still, while science continues to develop, or any sane person who obtains life and sustenance from our earth and sun. If it can be seen, spectrographed, measured, and defined, how can it be denied? It's real! It exists!

The Cosmic Observatory Background Explorer has provided us with answers that even Caesar could not reject. Science has all the cards and its messengers are technicians and observers of the universe. Unfortunately these new discoveries appear to fracture the unity of humankind and offer no explanation of Spirit, raising the question, can man live by science alone?

Parallel Linear GR: Introduction

In standard "big bang" models of cosmology, the structures seen in the universe today formed due to self-gravity; initially small perturbations in matter and radiation were gravitationally amplified into galaxies and clusters of galaxies. By following how these primordial fluctuations grow in time one can give testable predictions for the large-scale structure of the matter distribution.

Due to the same initial perturbations, there are slight deviations from a pure black-body spectrum in the cosmic microwave background (CMB) black-body radiation, a fossil relic of the Big Bang. For example, dense regions will begin to collapse, causing a rise in temperature; when-we look towards such a region, the CMB will appear slightly hotter in that direction. The COBE satellite, which confirmed the isotropy and black-body spectrum (with temperature 2.726 K) of the CMB, also measured these anisotropies. Subsequent balloon-borne and ground-based experiments have measured the amplitude of CMB anisotropies on different angular scales (see Steinhardt 1995, Lubin 1994, and Scott, Silk, and White 1995 for recent reviews). For a given cosmological model, one can compute the predicted amplitudes of the CMB anisotropies as another test of the theory.

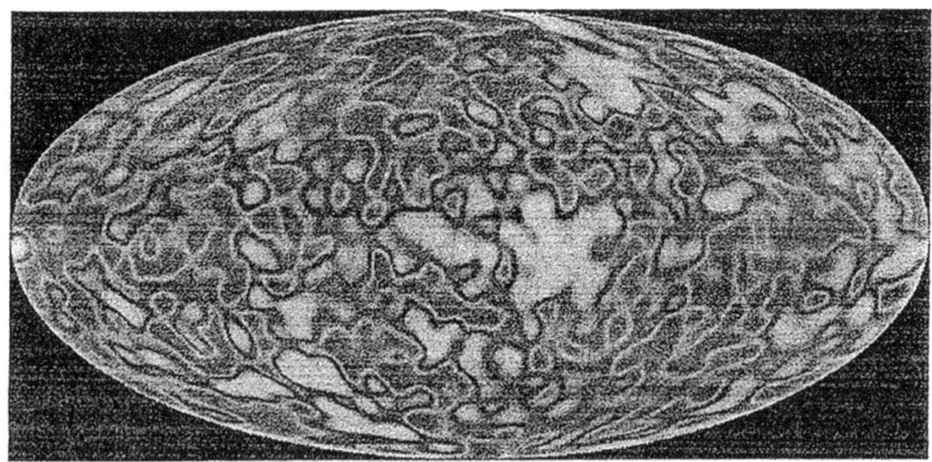

Structure in the COBE Differential Microwave Radiometer sky map (from the COBE home page; see Smoot et al. 1992 and Bennett et al. 1994). The DMR measures the difference in antenna temperature between regions of the sky separated by sixty degrees. Blue regions are cooler than average, and red regions are hotter. The map has been corrected to account for 1) the motion of the Earth with respect to the CMB, and 2) emission from the Galactic plane. The map has been smoothed with a Gaussian, which when convolved with the antenna beam results in a 10 degree smoothing on the sky. The variations are +/- 150 micro-K, relative to the average temperature of 2.726 K.

The CMB Anisotropies Measured by COBE

There are a variety of cosmological parameters which are not well constrained: the Hubble constant, neutrino masses, a possible cosmological constant, the initial perturbation spectrum, etc. However, once the initial conditions are set the input physics is well understood, and since the initial perturbations are small the subsequent evolution is accurately described as a linear system. The goal of cosmological spectral codes is to compute, for a given matter composition and initial spectrum of perturbations, the spectrum for mass perturbations and CMB anisotropies expected at the present time. These predictions can serve as a discriminant of the various models.

In the past varying degrees of approximation have been made in order to carry out the evolution (for example Peebles and Yu 1970, Bond and Efstathiou 1987, Holtzman 1989, Sugiyama and Gouda 1992, and Sugiyama 1995, among others). The code we discuss here has a highly accurate treatment of both the physics and the numerical integration; we believe it is the most accurate to date. The tradeoff for this accuracy is increased computational cost, making the use of super computers necessary.

The serial and parallel versions of the code (called LINGER and PLINGER, respectively) are available as part of the COSMICS (Cosmological Initial Conditions) software package.

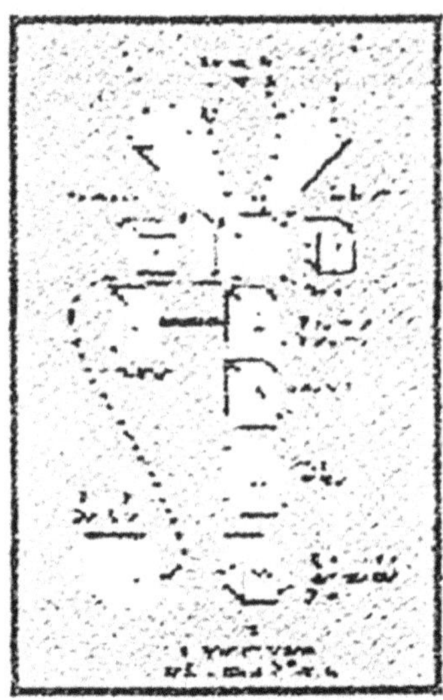

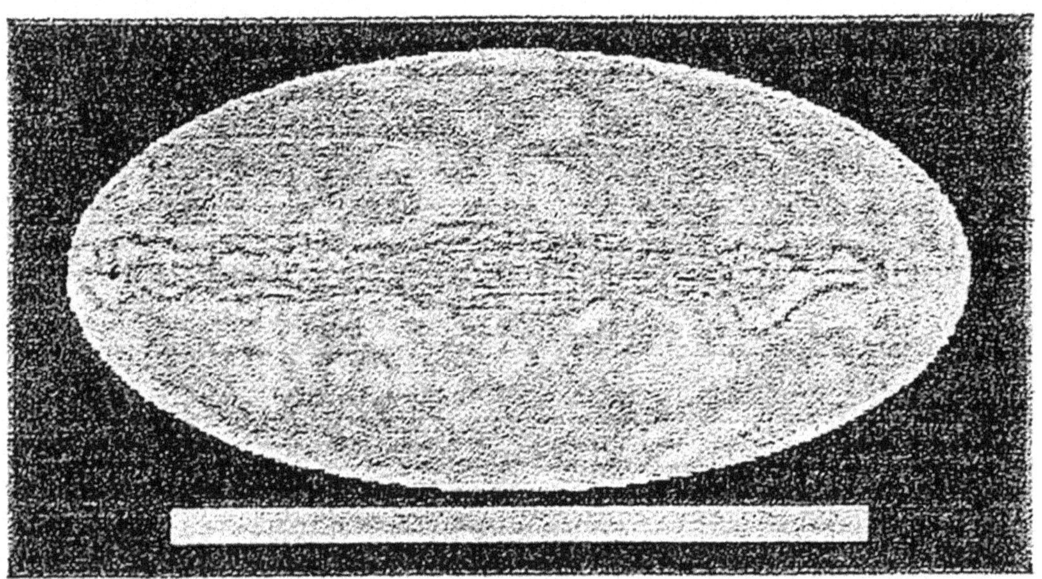

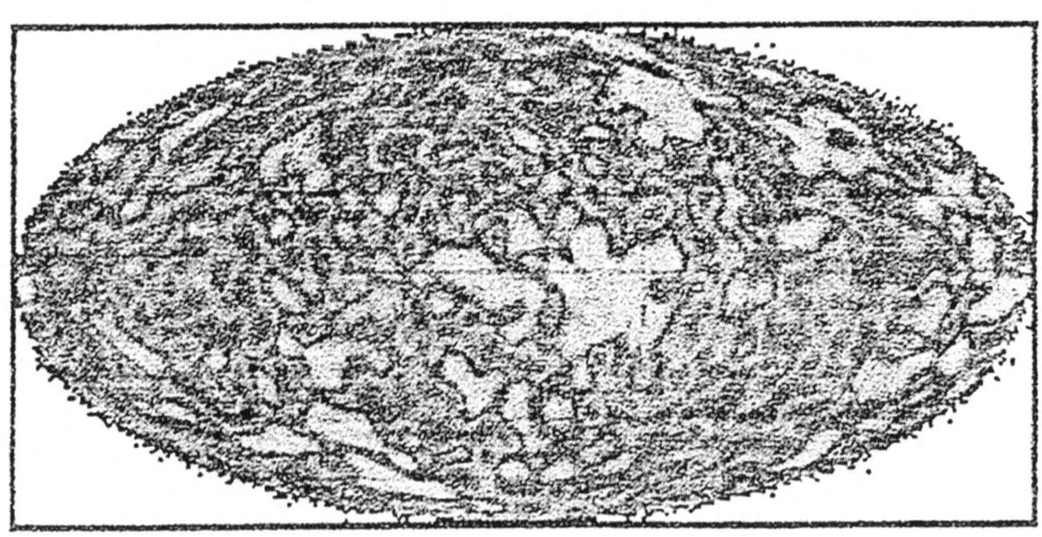

DMR Schematic

<u>DMR Four Year DMR images</u> of the Cosmic Microwave Background Fluctuations.

FIRAS (Far Infrared Absolute Spectrophotometer)

FIRAS has shown that the cosmic microwave background spectrum matches that of a blackbody of temperature 2.726 K with a precision of 0.03% of the peak intensity over a wavelength range 0.1 to 5 mm. Longer wavelength measurements, though not nearly as precise, conducted by Smoot's group at LBL and collaborators and by other groups, show that the CMB spectrum is well-described by a single temperature blackbody. See figure [attach Intensity and Temperature plots] of spectrum measurements. These measurements limit possible alternative models to the Big Bang extremely strongly and limit potential energy releases in the early Universe, typically to less than 0.1%to 0.01%.

- •CMB Intensity plot
- •CMB Temperature plot
- •Energy Release Limits plot
- •FIRAS Instrument Block Diagram

More Information

Extensive information about the COBE mission, including much information about the three experiments on board, is available by clicking on <u>more COBE information</u> or through the NTASA's NSSDC & Goddard Space Flight Center maintained <u>COBE Home Page</u>. More <u>photographs</u> related to and of COBE. NASA was responsible for the development of the COBE satellite and mission

FIRAS Scientific Results

The images shown below depict data taken with the Far Infrared Absolute Spectrophotometer (FIRAS) instrument aboard NASA's Cosmic Background Explorer (COBE). The colors generally do not map linearly into sky brightness. Use the FIRAS data products for quantitative analysis. Additional images are available in the COBE Slide Set.

To view the original images, click on thumbnails below:

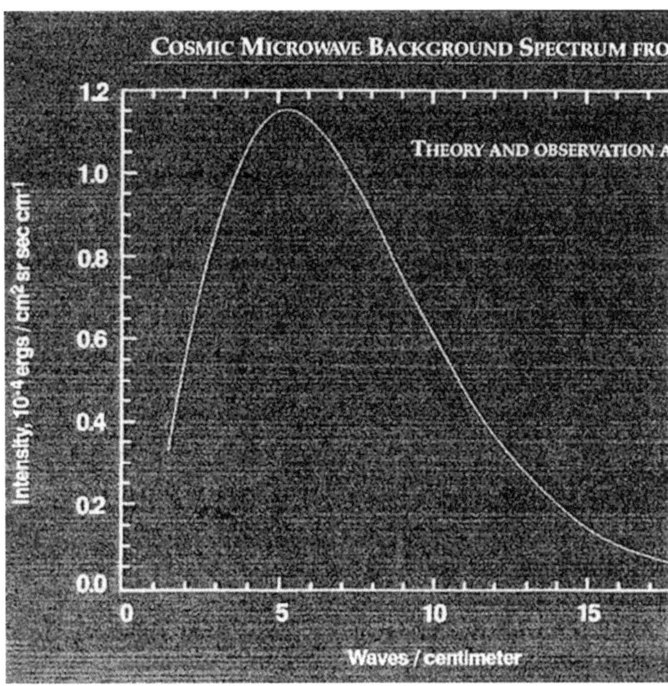

<u>Cosmic Microwave Background (CMB) spectrum</u> plotted in waves per centimeter vs. intensity. The solid curve shows the expected intensity from a single-temperature black-body spectrum, as predicted by the hot Big Bang theory. A

black body is a hypothetical body that absorbs all electro magnetic radiation falling on it and reflects none whatsoever. The FIRAS data were taken at 34 positions equally spaced along this curve. The FIRAS data match the curve so exactly, with error uncertainties less than the width of the black-body curve, that it is impossible to distinguish the data from the theoretical curve. These precise CMB measurements show that 99.97 percent of the radiant energy of the universe was released within the first year after the Big Bang itself. All theories that attempt to explain the origin of large-scale structure seen in the universe today must now conform to the constraints imposed by these measurements. The results show that the radiation matches the predictions of the hot Big Bang theory to an extraordinary degree. See Mather et al. 1994, Astrophysical Journal, 420, 439; "Measurement of the Cosmic Microwave Background Spectrum by the COBE FIRAS Instrument, Wright et al 1994, *Astrophysical Journal 420; 450*, "Interpretation of the COBE FIRAS CMBR Spectrum"; and Fixsen, et al 1996 *Astrophysical Journal 473: 576*, "The Cosmic Microwave Background Spectrum from the Full COBE FIRAS Data Sets" for details.

In addition to its primary, cosmological objective, the FIRAS provided important new information about the interstellar medium. The far-infrared continuum is formed by thermal emission from interstellar dust, while spectral lines are emitted by interstellar gas. Nine emission lines were detected in the FIRAS spectra: The 158 µm ground state transition of C+; the N+122 µm and 205µm transitions; the 37o µm and 609µm lines of neutral carbon; and the COJ=2-1, 3-2, 4-3, and 5-4 lines.

C+158 µm and N+2o5 µm line intensity maps from Bennett et al, 1994, *Astrophysical Journal, 434:587*, "Morphology of the Interstellar Cooling Lines Detected by COBE" (available electronically as an appendix of the FIRAS Explanatory Supplement). The maps are projections of the full sky in galactic coordinates. The plane of the Milky Way is horizontal in the middle of the map with the galactic center at the center. The C+ line *(top)* is an important coolant of the interstellar gas, in particular the "Cold Neutral Medium" (e.g., surfaces of star-forming molecular clouds). In contrast, the N+ line emission *(bottom)* arises entirely from the "Warm Ionized Medium" which surrounds hot stars.

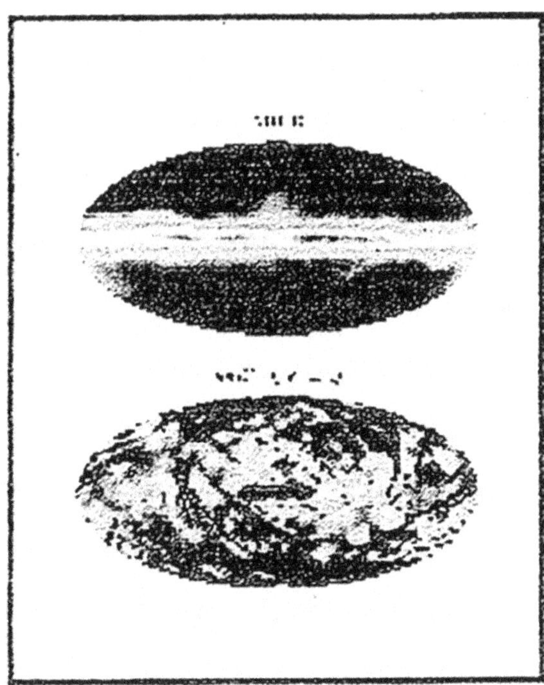

Maps of HI 21 cm line intensity and I(C + 158 μm)/N(H I) from Bennett et al. 1994., *Astrophysical Journal, 4:34*, 587, "Morphology of the Interstellar Cooling Lines Detected by COBE" (available electronically as an appendix of the FIRAS Explanatory Supplement).The projections are the same as those used in the preceding figure. *Top*: The distribution of atomic hydrogen smoothed to 10 degree resolution for comparison with the FIRAS data. *Bottom:* C + cooling rate per hydrogen atom.

DMR Images

The images below were created from the COBE DMR data products. Each image has been histogram-equalized, giving a nonlinear relation between color value and temperature. Use the data products for quantitative analysis. Additional images are available in the COBE Slide Set.

Before the full (four-year) DMR data set was calibrated and made into sky maps, preliminary maps based on 1-year and 2-year subsets of the data were created. While they have been superseded by the 4-year maps, these early renditions received much publicity.

Thus we feature on this page both 2-year and 4-year DMR maps.

There is an important qualitative difference between the 2-year and 4-year maps having to do with the improved signal-to-noise ratio that results when more data is analyzed at once. The 4-year maps, which include twice as much data as the 2-year maps, have a signal-to-noise ratio roughly the square root of 2, i.e., 1.4 times better than the 2-year maps. This modest improvement in map quality is sufficient to reveal individual features of the cosmic microwave background (CMB) in good contrast to the instrument noise. Due to the greater effect of instrument noise in the 2-year maps, those maps can be analyzed statistically to characterize the CMB fluctuations, but do not constitute accurate "pictures of the sky." However, in the final 4-year map, individual bright and faint spots represent actual CMB features. Illustrations of the major data analysis steps, and further explanations of the cosmic microwave background anisotropy (brightness fluctuations) are given below.

Maps Based on Two Years of DMR Observatior

Cosmic microwave background temperature data were extracted from the released FITS files and then combined into two linear combinations. The first is a weighted sum of the 53 and 90 GHz channels which gives the highest signal-to-noise ratio for cosmic temperature variations but includes the Milky Way Galaxy as well. In the second linear combination, a multiple of the 31 GHz map is subtracted from a weighted sum of the 53 plus 90 GHz channels to give a "reduced map" that gives zero response to the observed Galaxy, zero response to free-free emission; but full response to variations in the cosmic temperature.

These maps have been smoothed with a 7 degree beam, giving an effective angular resolution of 10 degrees. An all-sky image in Galactic coordinates is plotted using the equal-area Mollweide projection. The plane of the Milky Way Galaxy is horizontal across the middle of each picture. Sagittarius is in the center of the map, Orion is to the right and Cygnus is to the left.

The following image is similar to the June 1992 *Physics Today* cover picture, with the map including the dipole and Galaxy on the top, the dipole removed map in the middle, and the reduced map on the bottom. The dipole, a smooth variation between relatively hot and relatively cold areas from the upper right to the lower left, is due to the motion of the solar system relative to distant matter in the universe. The signals attributed to this variation are very small, only one thousandth the brightness of the sky.

Click on the image to view the original

The following image just shows the reduced map (i.e., both the dipole and Galactic emission subtracted). The cosmic microwave background fluctuations are extremely faint, only one part in 100,000 compared to the 2.73 degree Kelvin average temperature of the radiation field. The cosmic microwave background radiation is a remnant of the Big Bang and the fluctuations are the imprint of density contrast in the early universe. The density ripples are believed to have given rise to the structures that populate the universe today: clusters of galaxies and vast regions devoid of galaxies.

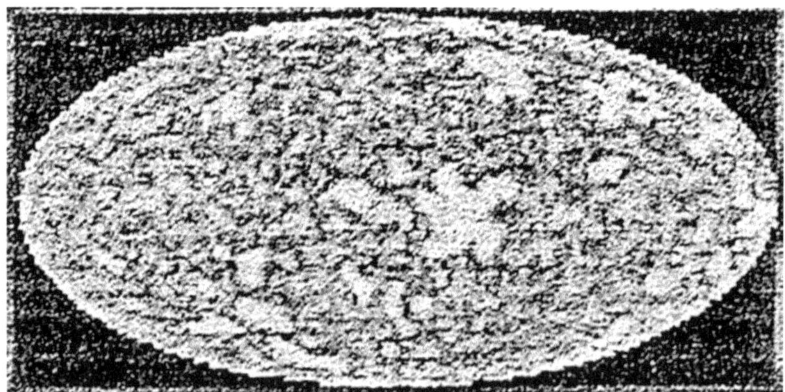

Click on the image to view the original

Maps Based on Observations Made Over the Entire 4-year Mission

Some of the features seen in the four-year maps below (i.e., dipole, Milky Way Galaxy) were described above. Like the two-year maps, the four-year maps have been smoothed to an effective angular resolution of 10 degrees. At this angular scale, the signal-to-noise ratio is sufficient (~2 per 10 degree patch) to portray for the first time an accurate visual impression of the cosmic microwave background (CMB) anisotropy.

The following image represents DMR data from the 53 GHz band *(top)* on a scale from 0-4 K, showing the near-uniformity of the CMB brightness, *(middle) on* a scale intended to enhance the contrast due to the dipole described above, and *(bottom)* following subtraction of the dipole component.

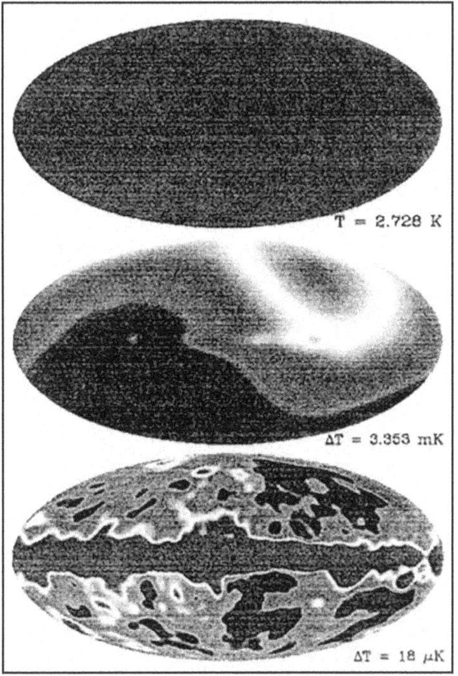

Click on the image to view the original

The next image shows data obtained at each of the three DMR frequencies—31.5, 53, and 90 GHz—following dipole subtraction.

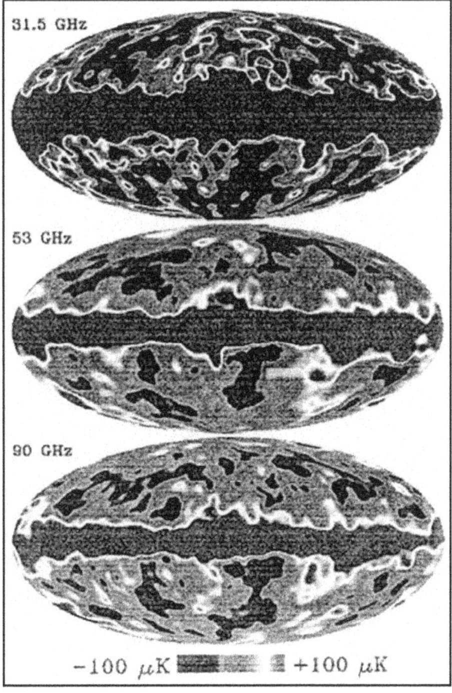

Click on the image to view the original

The following image shows the 53 GHz map *(top)* prior to dipole subtraction, *(middle)* after dipole subtraction, and *(bottom)* after subtraction of a model of the Galactic emission. The Galactic emission model is based on COBE/DIRBE far-infrared and Haslam *et al.* (1982) 408 MHz radio continuum observations (see Bennett *et al.* 1996, ApJ, 464, Li). Bennett *et al.* excluded an area around the Galactic plane referred to as the 'custom cut' region when they analyzed the CMB anisotropy. The excluded pixels are listed in two files that can be retrieved by anonymous ftp from the data/cobe/dmr/asd4cmbdata.gsfc.nasa.gov.

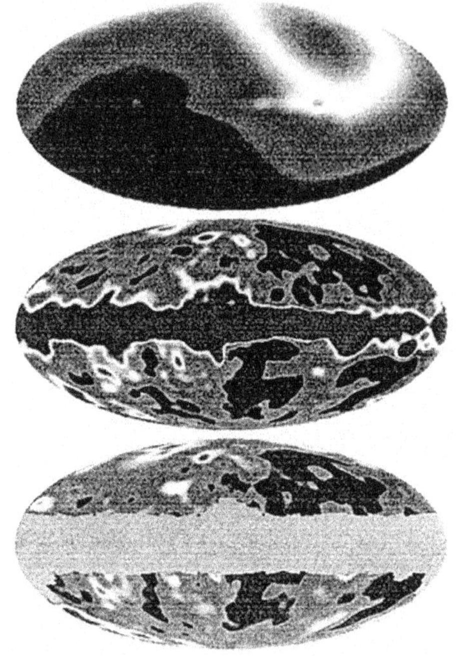

Click on the image to view the original

Finally, here is the COBE-DMR "Map of the Early Universe." This false-color image shows tiny variations in the intensity of the cosmic microwave background measured in four years of observations by the Differential Microwave Radio meters on NASA's Cosmic Background Explorer (COBE). The cosmic microwave background is widely believed to be a remnant of the Big Bang; the blue and red spots correspond to regions of greater or lesser density in the early universe. These "fossilized" relics record the distribution of matter and energy in the early universe before the matter became organized into stars and galaxies. While the initial discovery of variations in the intensity of the CMB (made by COBE in 1992) was based on a mathematical examination of the data, the new picture of the sky from the full four-year mission gives an accurate visual impression of the data. The features traced in this map stretch across the visible universe: the largest features seen by optical telescopes, such as the "Great Wall" of galaxies, would fit neatly within the smallest feature in this map. (Caption courtesy of Dr. Charles L. Bennett)

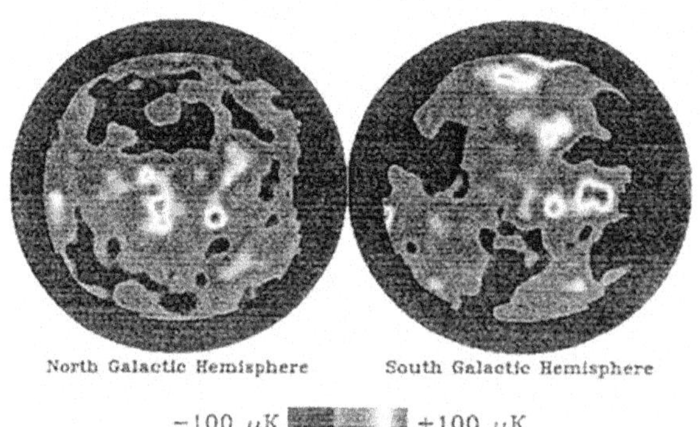

Click on the image to view the original

DIRBE Science Images

The images shown below were created from the COBE <u>DIRBE data products</u>. The colors generally do not map linearly into sky brightness. Use the original data products for quantitative analysis. Additional images are available in the <u>COBE Slide Set</u>.

To view the original images, click on the postage-stamp versions:

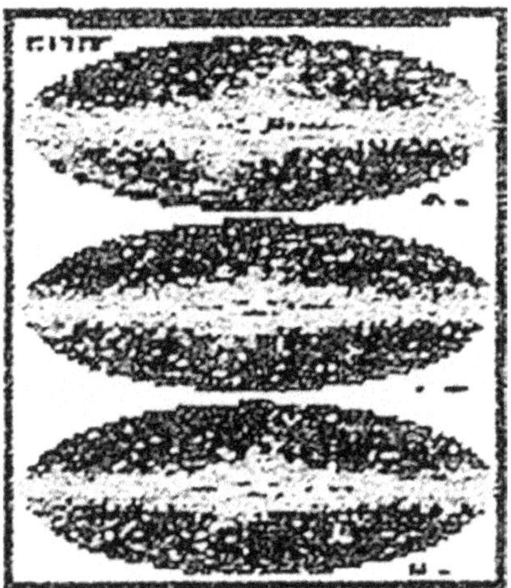

1.25, 2.2, and 3.5 micron Solar elongation angle=go deg Maps-Galactic coordinate Molhveide projection maps of the entire sky as seen by the DIRBE at a fixed angle relative to the sun. Stars concentrated in the galactic plane (horizontal feature) dominate the images at these wavelengths. Dust in the Milky Way absorbs and scatters starlight, producing the dark band that runs through the galactic center in the 1.25 micron image; this "extinction" effect diminishes with increasing wavelength.

False-color image of the near-infrared sky as seen by the DIRBE. Data at 1.25, 2.2, and 3.5m wavelengths are represented respectively as blue, green, and red colors. The image is presented in galactic coordinates, with the plane of the Milky Way galaxy horizontal across the middle and the galactic center at the center. The dominant sources of light at these wavelengths are stars within our galaxy. The image shows both the thin disk and central bulge populations of stars in our spiral galaxy. Our sun, much closer to us than any other star, lies in the disk (which is why the disk appears edge-on to us)at a distance of about 28,000 lightyears from the center. The image is redder in directions where there is more dust between the stars absorbing starlight from distant stars. This absorption is so strong at visible wavelengths that the central part of the Milky Way cannot be seen. DIRBE data will facilitate studies of the content, energetics, and large-scale structure of the Galaxy, as well as the nature and distribution of dust within the solar system. The data also will be studied for evidence of a faint, uniform infrared background, the residual radiation from the first stars and galaxies formed following the Big Bang. For more information about the Milky Way at wavelengths ranging across the entire electromagnetic spectrum, see the <u>Multiwavelength Milky Way</u> website.

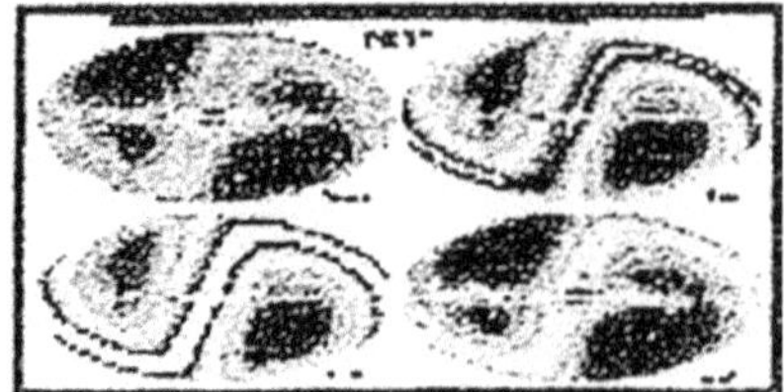

<u>4.9,12,25, and 6 omicron Solar elongation angle = 90 deg Maps</u>—Thermal emission from star-heated dust in the Milky Way and interplanetary dust heated by the Sun dominates the images at these wavelengths. The S-shaped feature is the ecliptic plane, in which, like the planets, the interplanetary dust is concentrated. The oval-shaped brightness discontinuity is an artifact of the way the maps were prepared, not a feature in the infrared sky. Specifically, the discontinuity corresponds to a path difference through the interplanetary dust cloud as adjacent positions in the sky were observed from DIRBE's vantage point in earth orbit with the earth on opposite sides of the sun.

<u>100, 140, and 240 micron Solar elongation angle=90 deg Maps</u>—Thermal emission from relatively cool interstellar dust warmed by stars in the Milky Way dominates at these wavelengths. At high galactic latitudes, interstellar "cirrus" clouds are apparent. Emission from the solar system dust ("zodiacal emission") is strongest a +25 microns but remains in evidence in the 100 micron image, and to lesser degree at the longer wavelengths.

All of the solar elongation angle=90 deg Maps are shown with logarithmic intensity scales. The following table gives the minimum and maximum log(I) for each DIRBE photometric band.

Minimum and maximum log(I) for DIRBE photometric bands

Wavelength (microns)	min log (I, MJy/sr) (black)	max log (I, MJy/sr) (whute)
1.25	-0.8	1.1
2.2	-0.95	1.3
3.5	-1.1	1.1
4.9	-0.5	0.75
12	1.0	1.7
25	1.2	1.91
60	0.7	1.91
100	0.42	2.0
140	0.05	2.71
240	0.05	2.71

The zodiacal and galactic emission must be precisely modeled and subtracted in order to detect the relatively faint Cosmic Infrared Background which the DIRBE was designed to find. Secondary DIRBE objectives include studies of these astrophysical foreground components.

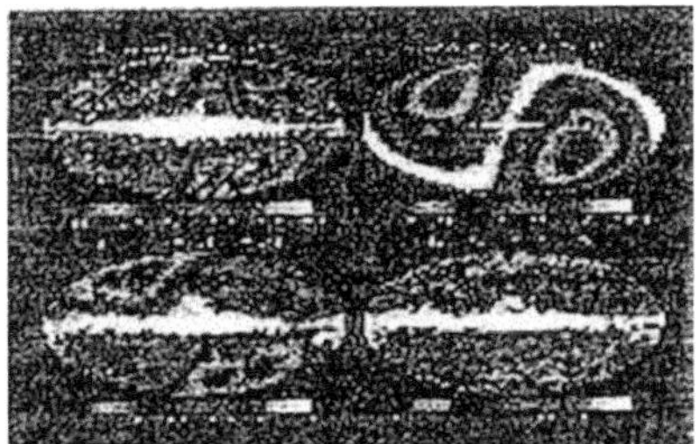

Annual average maps at 3.5, 25, 100, and 240 microns—galactic coordinate Mollweide projection maps of the entire sky at four wavelengths showing emission from stars and dust in the galactic plane (horizontal feature) and light scattered and emitted by dust in the solar system (S-shape).

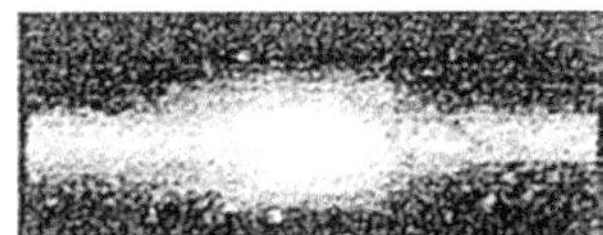

1.25, 2.2, 3.5 micron composite image of galactic center region—shows asymmetric shape of the bulge at the center of the Milky Way. The image is a Mollweide projection covering 60 deg in galactic longitude by 20 deg in galactic latitude and centered on the galactic center. A similar near-infrared image of the entire galactic plane is available (with zooming and panning features) on the *Multiwavelength Milky Way* page.

100 micron Weekly Sky Maps for mission weeks 4 to 44, plus annual average map—shows sky coverage each week of the DIRBE mission over the period during which the COBE cryogen supply lasted. As the Earth, with COBE in orbit, revolves around the sun, DIRBE viewed the sky from an ever-changing vantage point in the solar system, enabling light reflected and emitted by the interplanetary dust cloud to be modeled.

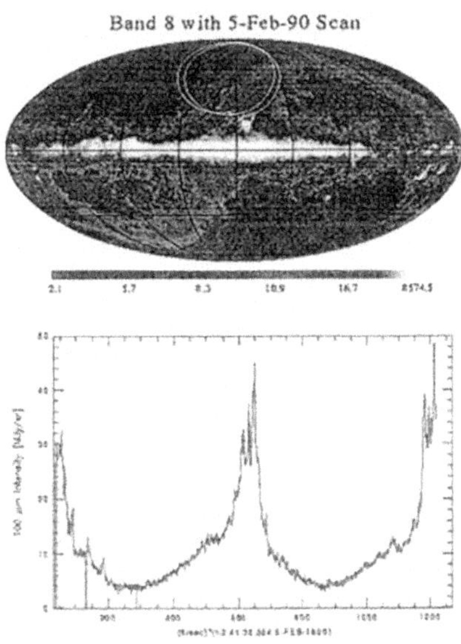

DIRBE scan track superposed on 100 micron Annual Average Map and 100 micron intensity from the corresponding segment of time-ordered data—DIRBE scanned the sky in a helical pattern that resulted from the spin and orbital motion of the COBE satellite and the "look direction" of the telescope, which was 30 degrees from the spin axis. The scan segment depicted covers two COBE spin cycles, or about 150 seconds, during a time when the DIRBE field of view swept through the Sco-Oph region (bright area above the Galactic center in the figure) and passed near the North Galactic Pole, where the emission is faint. The brightness of the sky at 100 microns measured during this interval is shown as a graph of intensity vs. time, as given in the DIRBE Time-ordered Data product. For the abscissa, the unit of time is 1/8th of a second. A related gif image shows intensity vs. time at four wavelengths—3.5, 25, 100, and 240 microns—during the same period. Zodiacal emission is evident at 25 and 100 microns; the bumps at about 470 and 1040 time units correspond to ecliptic plane crossings. Stars give rise to the spikes seen at 3.5 microns. The signal-to-noise ratio is clearly worse at 240 microns than at the other wavelengths.

The following two figures were provided by Dr. Henry T. Freudenreich and are described in his paper on *The Shape and Color of the Galactic Disk* (1996, ApJ: 468, 663). The Galactic plane runs horizontally through each figure. The Galactic center lies along the 0 degree meridian and the anti-center direction appears near the left side of each map. The 90 degree meridian is labeled for scale.

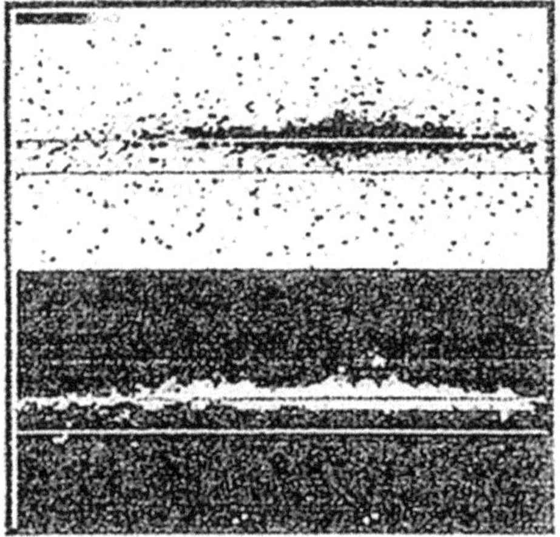

J-K and 240 Maps

Top: A sky map of the ratio of the surface brightness in the DIRBE J band (1.25 μm) to the surface brightness in the DIRBE K band (2.2 μm), after the zodiacal light has been subtracted. The range is 0.8 to 1.4. Starlight dominates the sky brightness at these wavelengths. The darker areas in this image exhibit the light scattering and absorbing effects of interstellar dust; those effects are greater at shorter wavelengths.

Bottom: The surface brightness at 240 μm. At this wavelength, in the far-infrared, stars are invisible and we see thermal emission from dust heated by starlight. The range of surface brightness is 0 to 115 MJy/sr. This map is bright where the top map is dark because the same interstellar dust clouds that absorb background starlight at near-infrared wavelengths are warmed by this absorption to about 20 degrees Kelvin, making them sources of far-infrared emission.

K-L and 240 μm Maps

<u>Top</u>: A sky map of the ratio of the surface brightness in the DIRBEK band (2.2 µm)to the surface brightness in the DIRBEL band (3.5 µm), after the zodiacal light has been subtracted. The range is 1.5 to 2.1. This map is darker where there is more dust along the line of sight, like the J-K map, but it also shows filamentar structure absent from the J-K map, a sign of dust emission at 3.5 µm, which requires a temperature of nearly 1000 Kelvins. An ultraviolet photon could heat a large molecule (e.g., a polycyclic aromatic hydrocarbon, or PAH), or a very small dust grain to such a temperature.

<u>Bottom</u>: The surface brightness at 240 µm multiplied by the sine of the Galactic latitude. This procedure accentuates the relatively nearby interstellar clouds, such as the molecular clouds in Orion (lower far left), Taurus (to the right of Orion), and Ophiuchus (above galactic center). When scaled by the sine of latitude, the map intensity ranges from 0 to 13 MJy/sr. Comparison of this map to the K-L map shows that regions containing cool dust tend to coincide with regions containing hot, small dust grains. The molecules or tiny grains seem to be a general component of the interstellar dust.

"THE PRIMEVAL GERM"

It is shown surrounded by the void of oblivion and displays the evolutionary concept of change that produces magnetic reaction induction or "ENERGY"

"You Cannot Save Yourself By Trying To Save Yourself."

Chapter 9
The Primeval Germ

In the search for origins, these masters of science have become robotic and the results, that they have provided us with are spiritless revelations of the beginning. They observe, they experiment, they make notes and come to machine like conclusions. It seems that the all of alls can be identified by science, but not totally explained by information that is based on data that was downloaded from spacecrafts and the supertelescopes and instruments that they carry. It appears that only religion can add the feel of a living spirit (intrafarance) in equating this cosmic background radiation with God.

There is a spirit that dwells deep within our souls, beyond the reach of the supertelescopes or microscopes or spectrophotometers and all man-created instruments. This is the undissolvable bond (intrafarance) that exists between living things; it connects them to the earth, the sun and the stars. It is an invisible organic umbilical cord, an inner awareness that links us with the source of life, the "Primeval Germ."

It is this source that we seek, that can only be rediscovered from within ourselves, by experiencing a new awakening that combines science and religion and technology and this new age of "Aquarius." This brings us back to the ancient Egyptian text that has provided us with the clues to finally understand the whole complete story of all of creation that depends on light or radiation for existence.

I existed in the primeval matter that evolved in nothingness, eons of evolutions ago, before the beginning of time. When I thought the moment had come, I fashioned my mouth and spoke my name as a word of power. With this power that was my birthright, I fashioned myself in form from the primeval matter of nothingness. I was alone at this time. There were no others with me then. I am the beginning of time, which began when I took form and commanded the power.

All that is, is because of me. The heavens and all things contained in the heavens are my works. I put the universe into being and stationed each and every star, planet, constellation, asteroid and thing of being. If they are to endure till the end of time, then in the order that I stationed them, they must remain. For it was my plan to station every heavenly body and all things in the order of existence.

When a star doesn't exist any longer, then it has left its station. I placed each and every star and thing where they are now found so that together they work one for the other and their movement is no drain on my power. In their proper stations the stars filled the darkness of space with light and the things and beings give meaning to my works."

Now it was that after I had come into being and had taken form, I found no place upon which I could stand. So straight forth from my thoughts, I created the primeval mound within the ocean of Nu. Being that I was alone at that time, it pleased me to make others. So I went from being one god to gods three. I created the god Shu and the goddess Tefnut. I made them in a manner that all things are made.

> I the creator of creation, the father and the mother, the he and the she of existence, had union with my member and hand. I embraced my shadow and I loved myself. From my discharge, I made the god Shu, then I gathered together the remains and swallowed them.
>
> I afterwards spat from my mouth a portion, and from this was made the goddess Tefnut. They are the gods of my heart and soul, for they come forth from myself and their forms I duplicated many times. Shu and Tefnut had union and from this union was born the god Seb and the goddess Nut [intrafarance]. And in similar fashion I duplicated their forms many times. Seb and Nut had union and from this union was born the gods Her-ur, Ausaures, Set, and the goddesses Ast and Neb-het.
>
> They were all brothers and sisters and born from one birth. From these gods and goddesses came forth mankind and things that creep and crawl, swim and fly, and all manifestation of things. And when all this was done the universe and all matter had come into being. I Ra-Nebertcher, am the Evolver of evolutions; the germ of primeval matter, the spirit and the soul of existence. It is I who issued the order of existence.

Each time that I reread this amazing text I find it more difficult to believe that it is an ancient text, that is possibly 2,000,000 years old. From this theory of creation was developed the belief in a superior being, the beginning of spiritual awareness, and the arts of science and religion, on this planet Seb Earth, Validating the fact that there actually is an order that applies to all things. If the mind can comprehend this, then it opens to the order of existence, G.U.T. or Grand Unification Theory. This is the spirit that is lacking in modem science and in religion.

This order of the light will provide the answers that are the mirrored secrets of the cosmic background radiation (or the cosmic egg..) This text describes a special ceremony that occurred in an ocean of darkness wherein the light was produced.

What is this ocean of darkness? That this sort of environment existed in the depths of outer space and is described in these writings is astounding! I began to imagine an environment of cold and hot dark antimatter filled with electrified subatomic and atomic particles, doing what it is they do to make light. What sort of place was the creator speaking of? A place where matter can be created-and all of a sudden it came to me. By applying the law of dualism, I came to understand and identify this ocean of darkness as a blackhole.

Under normal circumstances matter cannot be destroyed, but these massive oceans of darkness or blackholes do not equate to normal circumstances and possess the absorption power to make the light go away or disappear. It all began to fit. The whole of the entire universe began to actually make sense. The theme behind the creator of the light was real.

The process that he describes involving the makings of light could actually be related to the stars. If this much is clear to the reader, we can now go onward, secure in our understanding of how the light is developed and its true origin. Bringing us to the most substantial of questions: Where does the light travel to, inside a blackhole (the singularity)?

It is believed that either the singularity or oblivion awaits the light inside of a black hole, causing its doom. This is the ending place of stars and clusters of stars, where they are destroyed by these swirling orbs of dark radiation. Inside these blackholes lies eternity come to an end, and intrafarance is present.

Would the pope of Galileo's time comprehend this concept? Would the father of Aryan supremacy in his madness comprehend these thoughts? How would this message be received by the Chaldeans, Babylonians, Persians, Greeks, and the Caesars of mighty, Rome? Can the reader now understand? This book is meant to commemorate and announce that the age of Aquarius is upon us. It is a message that could not be delivered in the previous age of Pisces. The racist European doctrines of might and right, would not permit this.

This is, the story that reveals where the stars are born and how they can die. This is: "far out!." And as a descendant of this African culture that developed it, it fills my heart with elation. This "cosmic" "depth" is not recognizable in the ancient writings of any other earth culture.

It takes us now to a place where, sub-atomic particles are, natural emissions of the growth cycle of a space germ, that derived origin from an unseen place called "oblivion." It is the place that brings "light" and "radiation" to an end and allows the re-beginnings of origins. It is a source that invokes the laws of dualism and of eternal perpetuation (intrafarance).

Let us reflect on the big bang theory and its short comings. It does not identify oblivion! It does not explain the how or when or where or why there is existence. It is a theory without soul and it was developed in the waning hours of the past age. It does not include the complete laws of dualism. For instance, if our universe is expanding, it, suggest that somewhere our universe is diminishing.

This law of opposition must be present in explaining the beginning of the universe. Let us imagine and develop within our thoughts a spiral Microvian and Goliathian universe wherein you have expansion and subspansion. That universe would expand into, or diminish into oblivion. Could this be the case with our universe? If so, it seems to be doing so in an orderly fashion. Everything about it is orderly! Let us now dismiss this theory (big bang) and develop concepts that apply to oblivion.

The supertelescopes have provided the proof of uniform expansion taking place, confirming that regardless of the age of the universe it is still undergoing development. The process of creation is never-ending. If it were to stop, its ending would signal the start of a new beginning. This is the spiritual and religious message that is radiated to us, from the primeval germ that began existence within oblivion (intrafarance).

"OBLIVION"

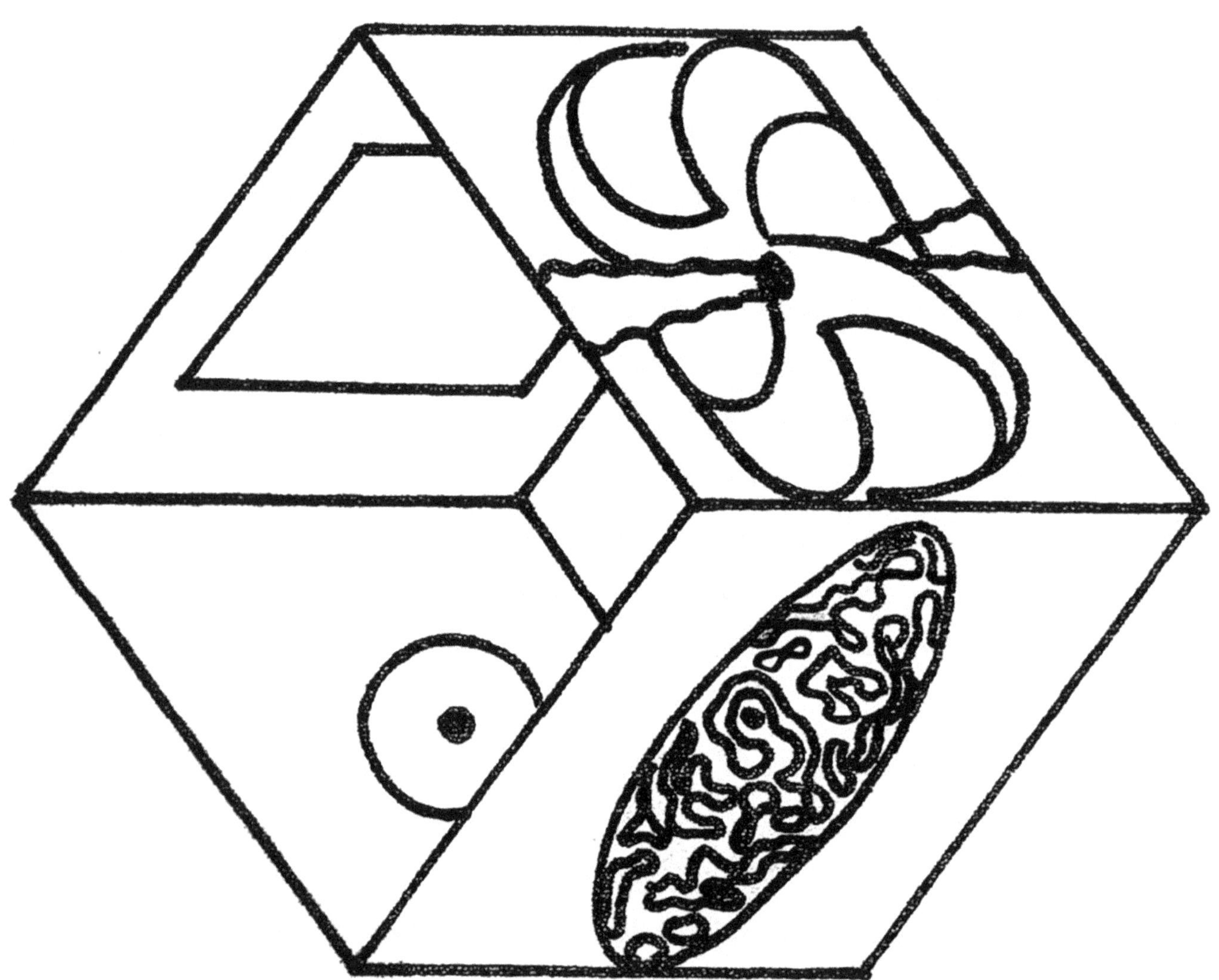

Is realized through the concept of reduction, reduction is the result of subspansion or expansion, The cube of existence is drawn into the delirium of "oblivion" where all it's energy forms are transformed and captured wherein the universe comes to an end.

"OBLIVION"

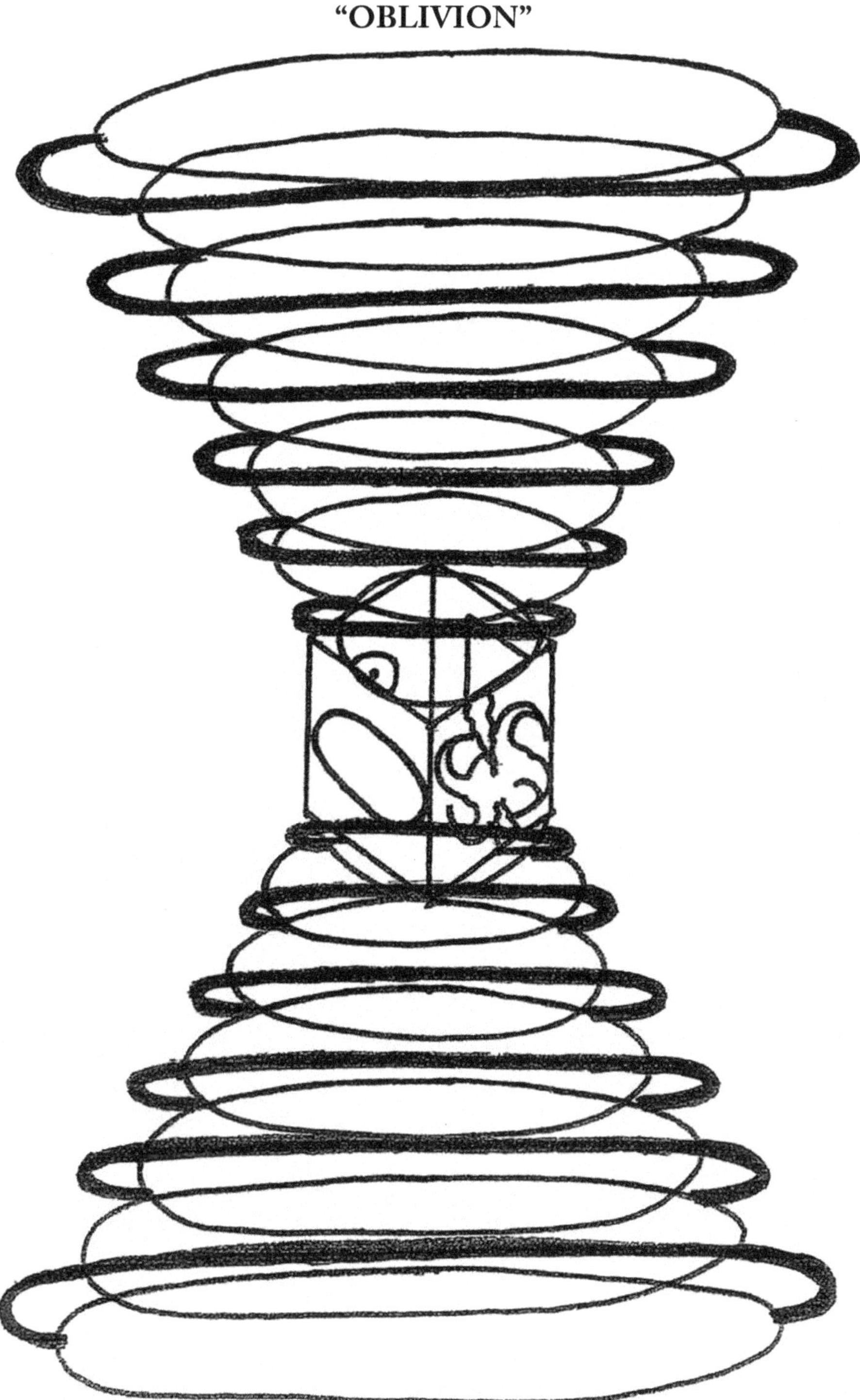

The collecting of particles that produce "existence" this illustration depicts "oblivion" as spiral rings that engulf the cube of existence.

The Wall (Racism) Must Fall, For Sake Of The All.

Chapter 10
Oblivion

Oblivion is a place of perpetual continuance, where particles of existence are reduced to their lowest common denominator (invisible electron strings). To understand this concept in full, we will need to apply the science of quantum mechanics. This is the science that allows for the disappearance of and there appearance of electrons, which many believe are (can be reduced to) the smallest particles in existence. They have the ability to appear and then somehow make themselves invisible. That is, they always seem to be present and they are everywhere; however, the ability to locate them at all times is as of now, impossible. If such a thing as time travel is possible, it may be that the electron is the clue. Quantum mechanics has shown that electrons have properties that allow them interdimensional behavior.

Oblivion is the place that allows this unusual behavior of electrons. It is a form of existence that connects the concepts of subspansion and expansion; its boundaries are endless and ever-reaching, indicating that it is the outer cover of the universe of light and radiation and that it is the inner cover that encloses this universe of dualism. Definitions that relate oblivion and intrafarance do not exist in the English dictionary. This definition of oblivion is also not to be found in the eg/dict.

This is the place of the "All of Alls"! The "holy of holiest" that sandwiches the universe of light and radiation. This qualifies it as the place or plain where matter is reduced and absorbed and then stored, for the purpose of continuous perpetuation, where the particles of light and radiation are collected for the redistribution process involved in continuous evolutions. This is how the ancient Egyptians equated oblivion (intrafarance).

Was this what was meant by the term "underworld"? It would appear to hold true with their doctrines of the light. The underworld was a place to which upon death the worthy would travel and await their rebirth, from whence they would be born anew as stars!

It seems to fit the description of the domain that has always been associated to or with Ausaures: a mythical place, where departed souls are born anew. I trust that readers can appreciate this ancient African version of the creation of everything and now begin to see their religion in a less complicated perspective. It is as real and complete as the stars are brilliant and it provides soul and spirit.

When Alexander, called the "Great," had subdued the Persian king Darius by defeating all his armies, he was presented with a challenge of might, right, and power. In the presence of the high priest of this ancient culture, Alexander was given a network of rope and knots, with a thousand ends. It was the ancient Persian equation of "all knowing." This menagerie of twisted hemp was called the "Gordian Knot." Its concept was unknown to the Greeks.

Alexander and his wise instructor Aristotle pondered the thought of its unraveling, and Alexander hesitated, for the first time in his young life. The mystery was beyond the comprehension of this young upstart and his learned advisor, so he unsheathed his sword and struck it. It fell upon itself and was no more. The high priest knew at that time that wisdom and right had given way to ignorance and might.

The priest revealed unto the young conqueror that from that time onward this would be "the way of the world." In the mind of that high priest, Alexander and his European school of thought (as instructed to him by Aristotle) was about to set the annals of cosmic understanding backwards. The world was about to experience the art of war. It was religion that became its first victim.

As we have seen, this story is so extensive and because it was fractured it was nearly lost to us forever. So as we have arrived at the moment of truth, I will attempt to explain the origin of the cosmic egg. If the universe had expanded and sub-spanned into oblivion, the process of evolutions and of continuous rebirths began anew.

This! is how it happens. That elusive electron that we heard about in quantum mechanics is collected here, as is everything else. Everything that existed within the universe is nowhere. This is why it is known as the "all of alls." All the power in the universe is collected into one place, oblivion. In modern dictionaries, oblivion is obscure and not an actual place.

A special electron or "Xlectron," now reduced to an invisible strand or string (not) possessing an electrical charge, frees itself from the all of alls. Engulfed by oblivion it begins to self-create. This is the true essence of intrafarance. It does not enjoy the feeling of being weightless, similar to the creator of the light, who says, "I found no place upon which I could stand. Attention is paid to the uncomfortable sensation of weightlessness. To counter this, It seeks It-Self out (plenum).

The creator of the light said: "I fashioned myself in form." This is what x lectron does, by fluctuations of its ends causing Intrafarance (heat and movement-radiation-and the altering of two like ends into electrically and magnetically charged opposites, thus creating a circle or orb like shaped being, of form and existence.

This form allowed it to emanate energy from Itself and as this energy was released, it pushed off Itself, causing repulsion. In this manner magnetism becomes the first complete force. These emanations of repulsion manifested themselves into subatomic dark cold matter and created the inner void of black space. (This is gravity in reverse) This was the very beginning of the early universe, a beginning, that mirrors the creation of the light.

As we can now understand the solidity in the structure of the light and its development, we can assume that a similar process occurred in the creation (development) of radiation. What we are now developing (re-discovering) is the beginning of the "Innerverse." The creator of the light says: "I was alone at this time, there were no others with me then." It would seem that the act of self-creation is best done alone.

The primeval germ was also alone at this point and in the fashion of self-creation, it made others from itself (Intrafarance). It germinated within this cosmic egg (form) and evolved. This was done by expansion (an explosion) and the shedding of its outer skin or shell, which was pushed away into the depths of inner space. Only to reach a distance away from the center and stop. Similar to the "Helio Pause."

These bits of pure energy were "Anti-Matter," discarded fragments of the shell that encased the primeval germ. Upon reaching the end of the distance of repulsion, these fragments of "Anti-Matter" (experienced gravity) and collapsed into orb forms and became Gamma blasts, creating a ring or Gamma belt that surrounded the primeval germ within the cosmic egg. (Inner-verse)

This (separation) explosion that occurred then, was the most powerful despising of energy ever in all, the universe. Thousands of millions of times more powerful than "Super Novas." Constructing the Gamma belt where explosions (they) dispense their energy inwards towards the primeval germ and this acts, as a photosynthesis, providing life and renewal of growth for the germ of life.

This is what is meant by the term "innerverse." The light does not exist there, as we understand it. Within this innerverse of subatomic emissions, where particles of repulsive energy are converted into a form of digestive radiation that is absorbed by the primeval germ allowing it continuous growth and perpetual evolutions. It cannibalized itself and sustained itself through expansion. The innerverse is a cold, black, dark, inner space of magnetic subatomic particles, bounded by the germ of existence and the gamma belt. (See Hakka.)

The universe of light is the result of the energy released outwards by these explosions of gamma blasts. The energy that escaped the innerverse by passing outward beyond the gamma ring contains small portions of escaped antimatter. Upon experiencing "intrafarance" and becoming sources of gravity, they gather in an area of space that we recognize as the "Nursery of the Stars". This constitutes the "outerverse" or space that is viewable through supertelescopes, the domain of the stars. These two sections of outer (stars) and an inner (radiation) complete the makeup of the whole, entire universe.

This idea is not known of in the English dictionary.

Ideally, the Earth's position and that of our Sun, is outward from the innerves of the Milky Way Galaxy, affording us a view that allows us to observe this "after glow" of creation. We are on the edge (of the outward expanse,) of the outerverse. Looking inward, we see our neighboring stars and our sister stars, and the seemingly eternal vastness of space. This dark, black void is filled with even more stars and unknown Constellations, Galaxies and clusters of never seen before stars. Near bye is the nursery of stars and the sub-atomic boundary that becomes noticeable as we approach the Gamma Belt. For the light begins to flicker and fade as we get closer. The reader with a very keen mind can envision (a) Bacteria, that is sustained by the requirements of an outer-verse and an inner-verse.

As we travel beyond the Milky Way inward towards the "after glow" where no telescope can see, beyond the operating distance of man's instruments, and approach the end of the atomic barrier (limit), where the light can no longer exist and only our spirits can continue, we arrive at the "Ring of Gamma" and our souls shudder in the awesome wonderment of power, that is displayed, unceasingly in this domain, that is not of, or for the light. Just one gamma blast of pure anti-matter is equal to all the supernovas that have ever been observed. Within this gamma ring, thousands upon thousands of explosions are occurring at regular intervals. They bear witness to the concept of photosynthesis. And in this way the energy of the innerverse is contained and managed!

This barrier allows us to give shape, form, and size to our seen and unseen universe. (In our model it is cubed and squared.) The afterglow or the primeval germ-the cosmic background radiation or the cosmic Egg-takes up two thirds of the space (59,000,000,000,000,000,000,000 miles) within this universe. From its center, the innermost part, to its outer, fractured, fragmented, and living shell, called the gamma ring or belt that surrounds it, is a distance that is equal to two thirds of known space. Which indicates that subatomic space (there are no stars) is twice as large as atomic (stars, light) space. The remaining third, that includes our Milky Way and all the other stars, exists outside of this reverse gravity source of radiation-a life form that has shape (containment) heat, movement with the absence of gravity-that allows the expansion of itself, the gamma belt, and the remainder of the universe.

Every natural function in the universe must relate to this (cosmic background radiation) outer-space germ of life that is the primary source of radioactive creativity in the early universe. Theories of origin are now without meaning, unless they reveal these aspects of light and radiation and of heat and movement that collectively announce the order of existence. Light requires gravity! C.B.R. is the essence of gravity in reverse. This is why G.U.T. equations break down inside of black holes, as do light atoms. This allows new definition to terms, such as phenomena (gravity) and anomaly (absence of gravity).

Could these be the pieces of the puzzle that N.A.S.A. is lacking? For none of this would make sense if not for C.O.B.E.

These little notes equal the thought and order that can become a quick saturation of thought! We travel beyond the door of European understanding!

Part II. The Inner Book

"HAKKA"

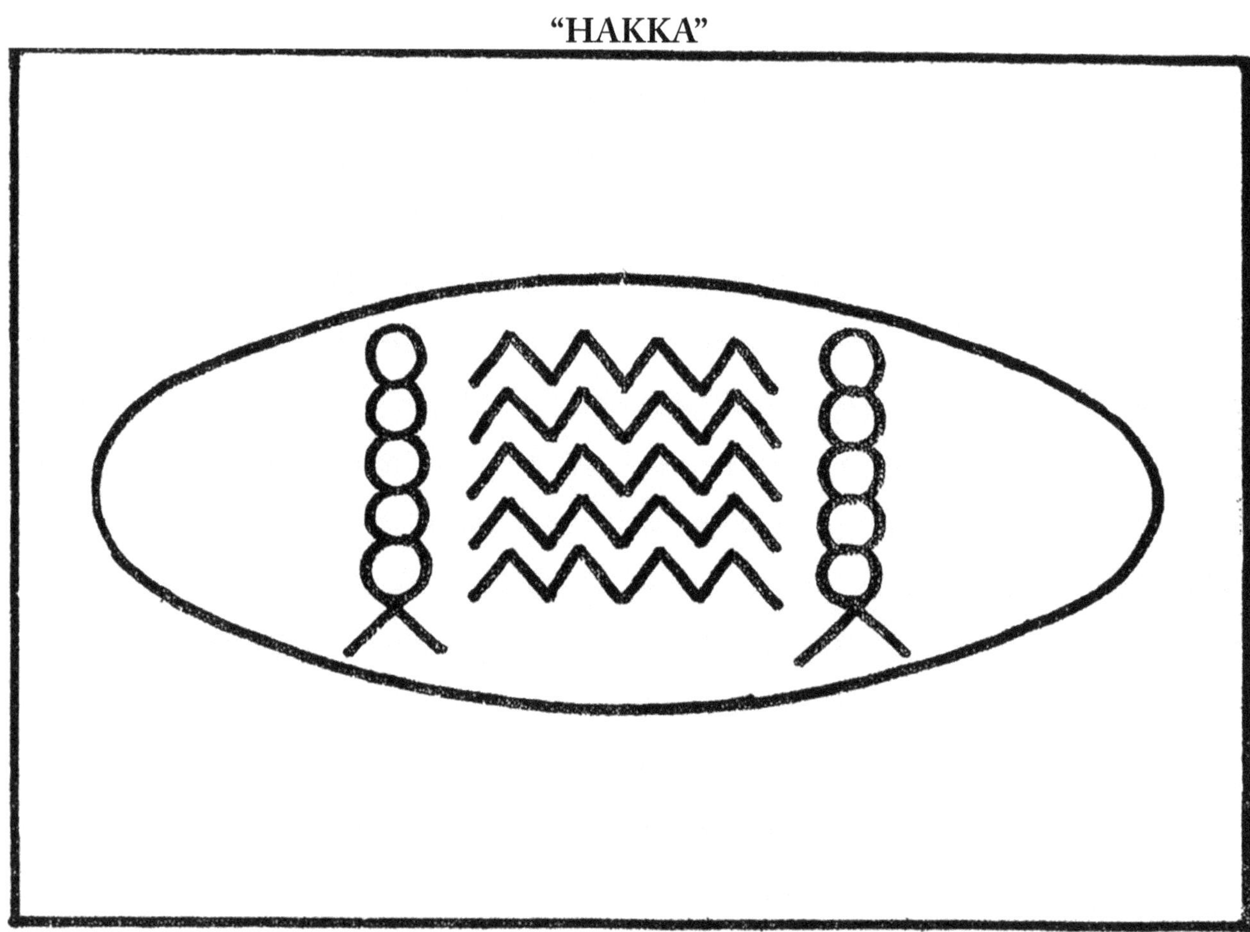

The magic that makes the Universe work. This concept was rejected by the Greeks and Romans and its likeness (cartouche) was purposely destroyed, its understanding was soon lost.

Chapter 11
Hakka

The ideas in this book concerning the creation theme were expressed in one word in ancient Egypt, *Hakka*. I trust that it will add to the reader's understanding of this book. The singularity, the all, the one, the magic that makes all of the universe to work was called "Hakka!"

I have updated its retranslating so that now it describes a substance that was unknown in the early years of the eighteenth century. This substance is radiation. While most scientist and scholars, then and now, consider magic to be the art of trickery, radiation is a magical anomaly. It cannot be completely defined or understood. However, it can be explained in part and understood in part. To comprehend these parts requires knowledge of the atom.

All the particles of ethereal existence (matter) are made of atoms. The atom has weight, shape, and structure. The universe of light as we understand it owes its existence to the atom. In this regard the atom: (phenomenon) It is the singularity, the All that is found in the One. It is reflective of the magical forces that brighten our universe of light. In this very active state, magic is real, it does exist. It exists within the atom and all atoms. It is the power to change, rearrange, and alter. This is the very essence of supernatural power.

There is magic, and the atom bears witness to this. However, its magical powers have a limit: that undisturbed, that Atom alone cannot un-become and then become again. It is a wondrous source of power, but it cannot create itself. It is not power complete. To the ancient Egyptians, hakka" was the (spark of) energy that existed before the atom. Hakka (intrafarance) Existed before there was a before. To explain this we must decimate the atom and arrive at a state of existence before the light and its atoms came into being.

All atoms have electrons that revolve around a nucleus that contains a proton and a neutron. When the electron is dislodged from its orbit and moves inward towards the nucleus, the atom will begin to radiate, expelling energy. As the electron moves away from the center, the atom will absorb energy. (This could be the universal law of assimilation and decimation of atoms and radiation-black holes.)

Absorption creates heat and radiating produces (outward) movement. In order to experience life there must be heat and movement. Therefore radiation is a necessity in the cosmic order of life, as are atoms, necessary in the development of the light. They share a common benefactor and that is the electron.

The belief in a creator of light (or matter) is not the indulgence of magical fantasy. It is atomic fact. If radiation is older than light, then its particle structure must be subatomic. This is what is meant by the idea of light and radiation being mirror images of each other. The art of science known as physics was first developed in the pursuit of knowledge concerning the light. Modern physics is the study of radiation-rays, waves, and their relationship to light. (matter). This knowledge was known to the ancient Egyptians Hakka was the understanding of these magical forces of creation.

Astronomy today is derived from cultures that existed in ancient Asia, premodern and modern Europe, and modern day America, both North and South. It is the study of lights that appear in our night skies. If this twilight art of observing the stars existed before the telescope, then why are Africa's contributions to this subject ignored? Hakka is the result of these African minds that for many more centuries have been aware of the cosmic wonder of wonders.

The creator of the light describes an existence in an ocean of primeval matter. What is this primeval matter? Is it something that was known to these so-called fathers of astronomy who that were spawned in Asia, Europe, or America? I don't think so. It would take the invention of the supertelescope before they would spot this phenomenon. of primeval matter (black holes).

As we have come to understand, these "Oceans" in the depths of space are black holes. Today we have many vehicles of the stars that have confirmed this. They have also rediscovered the cosmic egg, or the cosmic background radiation. Is there such a thing as astronomy with religion or just astronomy without it?

What is this cosmic background radiation? It is an "egg-shaped (reverse gravity=absence of downward-pulling) anomaly" that was located by the special equipment carried aboard that spacecraft, C.O.B.E. It is the largest structure in the entire universe, consuming two thirds of its volume. Its length is measured at 59,000,000,000,000,000,000,000 miles. The width is about half of that in the center, decreasing as it extends throughout the majority of space that is available. It is also the oldest anomaly in the universe. Its signature of red radiation is highly suggestive of it being God or the source that produced God.

I wanted to know: What is the origin of both of these cosmic forces that are contributors to life and order? Ancient astronomy as it was known outside the continent of Africa could not provide these answers. And with the advent of C.O.B.E. and supertelescopes and other spacecraft that carry similar instruments of observation, still, they are unable to provide these answers after ten years of constant, laborious study. Was I asking questions that are unanswerable?

By connecting the ancient Egyptian Creator to the black hole, I slowly began to realize that it held the clues needed to provide the answers to those two questions. I rethought the principles of matter and realized that it can be destroyed. Black holes have (forever) been creating and destroying matter as a natural continuation of the cycle of universal life. They posses the power to make light go away! This then must also be fact-that if it can be created, it can be destroyed. Nothing can endure within its dark swirl of Antimatter and subatomic particles. It is only through magic (hakka) that the light escapes.

It is known that black holes contain back body radiation that is visible at the infrared end of the spectrograph. Spectroscopy, the modern theory that involves the study of the atomic structure of atoms, has revealed to us that black holes are the interaction of light (matter) and radiation. The process in which the emissions of radiation and subatomic particles create light and the action of this matter being absorbed by falling on or into it (appears to) destroy matter. Dr. Stephen Hawkins has found that even black holes will dissipate due to loss of heat and movement.

This startling revelation indicated to me that both light and radiation are fallible. There is a place where they both journey and then cease to exist. I tried, inside my mind, to think of what (non) Existence is or was. I visualized within my thoughts the disassembling of atoms that allows them to be absorbed and changed into subatomic particles, which in their turn are also subjected to absorption.

The question must be asked: Where do they go? They go into oblivion. Up until that time, oblivion was a word to me, more than an actual place. Finally I had come to the understanding of what it actually was and is. A place, where light and radiation are collected and stored for the purpose of continued evolutions. It then dawned on me that if light and radiation are mirrored reflections of each other and are subject to similar endings, why would this not also apply to beginnings?

I was beginning to understand the order of existence in reverse. I needed only to have a greater understanding of this whole universal picture and then reverse it and arrive at the beginning. I was aware that the gamma blast, occurring at the farthest end of the remaining one third that was left in the universe, suggested a barrier that separated the two unequal sections. And that if my conclusions were to have merit, they must offer definitive explanations of this indifference. After feeling comfortable with my calculations I developed this theory on how oblivion can produce the "restart" of life and was a part in the initial order of the creating of (life) the universe.

I imagined that oblivion was an all, sided (cube), top and bottom, with a spiral of expanding and diminishing rings that enclosed the universe of radiation and light. From quantum mechanics I saw practical application of the theory of disappearing electrons and with the inclusion of the string theory it all became clear to me. Imagine, if you will, that all the energy of the universe was (is) collected into this place of oblivion. Taking the concept of the ancient Egyptians, that an end is a beginning and a beginning is an end, this was making sense. Life would just jump-start itself all over again. This is how that Jump-start occurred.

A harmonic string or strand was able to free itself or was released from the void of oblivion. And in the manner of the creator of light, it began the process of self-creation. This involves Intrafarance, the process by which that harmonic string became electrically charged, one end being a plus charge and the other negative. This allowed the string to seek out itself, take form, and overcome weightlessness. This string now was a living organism, possessing form, shape, and intelligence. It had become the primeval germ of radiation and life.

It straightforth began to discharge energy in the form of subatomic particles (magnetic repulsion) and the darkness (void) of early inner space was created. This indicates that the first universal force was a magnetic response to overcoming weightlessness. After taking form it emanated waves and rays of repulsion (radiation). These waves are still visible as C.O.B.E. has demonstrated and are the primary reason that the universe is moving outward (apart) from this germ of life or cosmic background radiation.

We have determined its approximate length and width, as it is viewable to us today. I imagined at its inception a germ particle that is microscopic, but needs to feed and grow. It evolved and outgrew its encasing or shell, and it expanded (exploded). These discharged bits of shell are antimatter, possessing the pure radioactive energy of this self-created germ. They were repelled outwards until reaching the barrier of magnetic flux or a helio pause, and then became a ring of energy that allows photosynthesis through gamma explosions or blast.

These explosions occurred because when they reached this magnetic barrier and developed their own gravity they collapsed on themselves and took orb form, resulting in the releasing of this enormous energy that is produced from antimatter (explosions). This is the barrier that completes the innerverse, that is absent of the light.

As a result of these gamma blasts, which are constantly occurring, small portions of antimatter escaped the innerverse and proceeded outwards, where gravity allowed them to become the seeding energy of the universe of light and today this area of visible space is known as the nursery of the stars. This is what was meant by the word *hakka*. The reason why our universe is expanding and in orderly fashion!

"INTRAFARANCE"

The spirit of change or the cube that endures "Oblivion"

Chapter 12
Intrafarance

Intrafarance is the principle or the spirit that is the one constant, that effects change in an ever-changing universe. It is the spirit of continuation and development. It is the force that causes change, or alteration. With its presence, the universe is forced to change, rearrange, and alter. It is unceasing, unalterable, and always at work. Water, which exists in three forms—gas, liquid, and solid—is a element that bears witness to intrafarance.

Changes in nature occur at random, caused by a storm, drought, or environmental relapse. Intrafarance is all of these things. It is the reason for the colors of light. The reason that radiation, radiates. It is the spirit of heat and movement, the master of death. It will not die. There is no force in the universe that is older than this. Before radiation there was oblivion; the principle of intrafarance existed. It is the spirit of beginnings and endings. This spirit cannot be isolated from anything. It has something to do with everything.

The other chapters of this book mention and introduce the word and concept. Here is a more in-depth examination of this Spirit of life and death. If we consider the evolutions of the universe as this book presents it, I trust that it is possible to understand this theory. Intrafarance is the theory behind theories.

I have prepared a graph that shows the effect of intrafarance on subatomic particles and its severance of changing them into atomic particles. It can effect change in any shape, form, or fashion. If we hold true to this text, we understand that a strand that possessed harmonic fluctuation was able to free itself from oblivion. Upon separation it underwent harmonic change. Soon it developed a harmonic rhythm with opposite charges. In both instances, intrafarance was the unseen spirit or principle. There is nothing that is, was, or will be that is not subjected to this living spirit of perpetually changing existence.

Intrafarance was not detected by the instruments that C.O.B.E. carried on board. It was not recognized, by the ancient Chaldeans, Babylonians, Greeks, or Romans. The scientists, technicians, and scholars who analyzed the downloaded data from C.O.B.E. and realized that they were witness to "after birth of Creation" also failed to see this magical spirit that was present everywhere. They had actually found it, the everything, the leftover of the Big Bang. What to name it was as difficult as explaining it. So they called it the cosmic background radiation.

I refer to it as the primeval germ. We observed the same object, analyzed the same data, and arrived at a different conclusion. On mostly all of their findings, I am in accordance with them. But how can you see the forest if the trees are in your way? Instruments and machines cannot comprehend the concept of spirit. Every human being on the planet is endowed with an insight that is far sharper and more microscopic than any of man's instruments, that would search for and then find and try to define spirit.

Only a living things can experience this unseen soul, this feeling of spirit that is so very real within ourselves. Intrafarance is the cause and effect that produces the concept of Hope and of knowing that there is some Thing that is God or Creator or the Maker that loves us all equally. Intrafarance is the pre-primeval spirit, the "Three in One" that extends to us all, the ability to change, rearrange, and alter ourselves, allowing each individual the freedom of choice. I suspect that if we could combine all of the history of humankind into one book-and let us call this the Book of Truth-it would be a revelation in equality. If we combine the science and the religion and the Space Age technology with meaningful knowledge from the past and add Intrafarance we have found the very existence of "God."

"THE LORD CREATOR OF THE LIGHT"

Ra Nerbertecher… I existed with the Ocean of Darkness and Primeval Matter" this illustration is realization of the Ocean of Darkness or "Black Hole"

Careful observation allows us to follow those steps that result in the emergence of the Light from within this "Black Hole" When he thought the moment had come, The right arm touches the head (symbolic of Thought.) The inner of the three legs is the partial body of "PTAH" the door that opened in the Darkness, the remaining legs are those of "AMEN" the "Hidden One" that Emerges, and the symbol of Khepara clutching the Solar Disk indicating that the Universe of Light is being fashioned, And these five aspects (combined) The encircling "Black Hole" where the ingredients are found that produced the Light. The two hands clasp the Specter of Light, that crowns with the encircling of the OLD RELIGION (O.R.) The wings on the right represent the colors of white light, while the left is symbolic of the Universal constantly changing color spectrum. The Radiation of Light emanates and outer space where atoms exists is realized

Chapter 13
The Innerverse and the Outerverse of the Universe.
(the Rausaurian Star Theorem)

This is the story of "universal dualism." The key players are radiation and light. Within the universe these two supply It with heat and movement. Without the understanding of these forces, the universe is without explanation. Radiation and light are the forces that allow uniform expansion and are the reasons the universe is expanding. They are mirrored images of each other, different only in their limitations, and from these forces come the laws of dualism.

Having an ending automatically suggest that there must be a beginning. For the end of one thing is where the other thing begins. A beginning is suggestive of an ending; one is the product of the other. This is how the universe must be perceived. "Where you find One you will also find the Other."

Let us consider the workings of a spiral universe, in which the laws of dualism dictate that expansion and subspansion, occur simultaneously. Theoretically both movements would result in oblivion. There they would come to an end and a new beginning.

Oblivion is the place where a universe comes to an end and where it also begins. The universe in which we live is subjected to these same rules. Once it was thought that matter could not be destroyed. Today it is known that black holes can destroy matter. They are the doors that (open) lead to oblivion. What are black holes?

Black holes are dark (radiation) altered "AntiMatter" that escaped in gamma bursts by moving outward of the gamma ring. They are the very first objects to demonstrate gravity in the outer universe. They consist of black-body radiation that is visible at the infrared end of the spectroscope. Within these black holes, there is an interaction of light matter and radiation. The process in which the emissions of radiation by matter create light and the action of this, matter being absorbed by falling on or into it, appears to destroy matter (light). Dr. Stephen Hawkins has demonstrated how and why these massive swirls of antimatter lose radiation and then they also disappear into oblivion. If a black hole consumes beyond its capacitance it will cease to exist.

Modern physics owes its development to the study of radioactive bodies. However, it was unable to completely understand what radiation is. A new school of thought arose, quantum mechanics, that governs phenomena which are so small that they cannot be described in casual terms. It is atomic physics, a way to calculate intensities and frequencies of spectral lines or heat emissions. The observance of the spin of electrons, with the energy of repulsion and attraction, explains in general the way atoms are held together. It is a field that studies the dispersion, absorption, and emissions of radiation.

The cosmic background radiation is the oldest and largest structure of existence. It will be known to the reader as the cosmic egg (aka the Primeval Germ), the original source of heat and movement.

We have seen how matter and radiation disappear into oblivion. Now let us understand how it is also released from oblivion. R. A. Schwaller De Lubicz writes in *Sacred Science* that: "Indeed, the "Beginning," the very "First" manifestation, is the "Essence" divisible into "Two" aspects." Energy that was collected and stored within oblivion is released. It is in the form of a harmonic or fluctuating string or a strand that possesses the ability to self-create.

"THE BEGINNING" OR A CONCEPT OF THE "BIG BANG"

That starts in Oblivion.

Dots and fluctuating strings, lines Intrafarance realized and energy escapes "OBLIVION" the C.B.R./Cosmic Egg/Primeval Germ is developed (self-creation.) it expands in an explosion of itself and Gamma Ring, is the result creating the "Inner verse."

From the field of quantum mechanics comes this string (strand) theory. It implies that the smallest or tiniest of molecules exists in this undeveloped (subatomic) form. This was demonstrated by the supermicroscopes that can magnify electron particles downwards, until they become strings. This is how we will first see and understand the development of the primeval germ. By relating the doctrines of the light and its concept of creation we will duplicate or copy a cycle of growth that is dependent on heat and movement. This string or strand that was released or escaped oblivion is called Xlectron.

Xlectron is alone at this time and upon freeing itself, it instantly developed needs that are responsible for the beginning of the creating of the universe. It was weightless, a "feeling" that had to be overcome. So it sought out itself, took form, and began to emit repulsive energy. These magnetic lines of flux allowed form and stabilization, and the effect of weightlessness was overcome.

It germinated inside its (new shape) egg and evolved. The cycles of evolutions had begun again! It needed to feed and there was only itself. But with the power to self-create, this was no problem. With the first ever (nuclear) explosion, it discarded the outer shell, released bits of antimatter, and doubled in size. As it expands so does the Universe.

The antimatter that was ejected in the blast was then repelled to a distance of magnetic pause, or stop, where it experienced gravity and began to take form (orb) and shapes, collapsing on itself and then exploding. These were gamma blast. The ignition of pure discarded particles of shell fragments or antimatter (pure radiation) sent energy surging inward and replenishing the hungry germ. It is similar to our photosynthesis. The cycle that allows perpetual evolution starts here. The boundaries of space began here, the existence of subatomic particles begins here, the laws of magnetic flux (attraction + repulsion) begin here, and the production of antimatter begins here. These are the facets of self-creation.

THE COMPLETE CUBE OF EXISTENCE

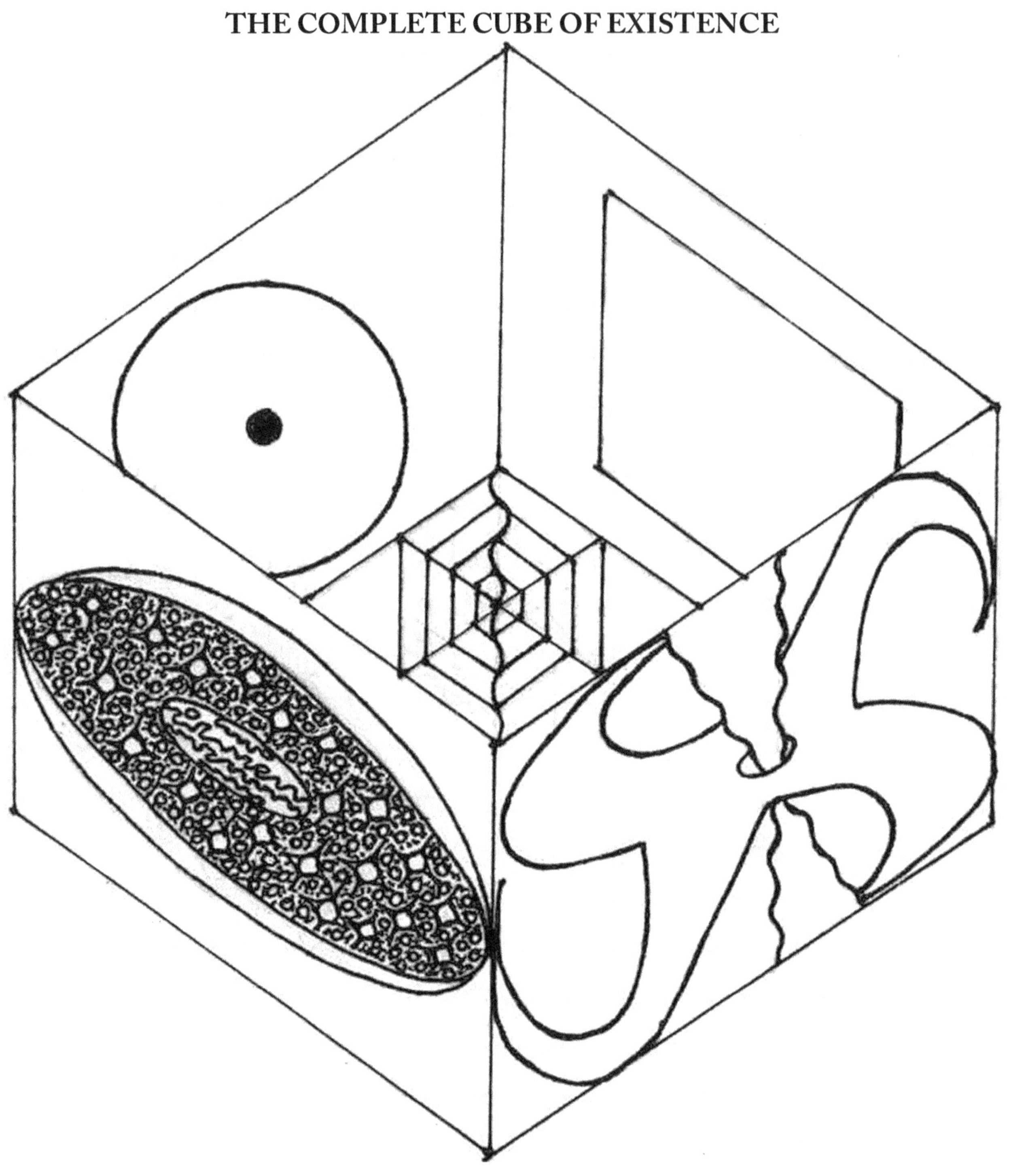

Chapter 14
The Gamma Belt

Gamma explosions produce rays that are electromagnetic radiations of very high frequency, emitted by bodies or particles of antimatter. They are the shortest waves of the electromagnetic wave spectrum. They appear similar to and possess the same ethereal properties as x-rays, except that they are of a higher frequency and much more energized. From the gamma ring that surrounds the primeval germ comes the term, innerverse. It consists of three parts, the primeval germ (egg,), its emissions (subatomic particles), and the gamma ring. They are the (eternal) fuel that powers this engine that is in the heart of space (a germ), an engine that is alive and still growing. There is no light (atoms) to be found within the innerverse.

Gamma orbs are the antimatter of evolutionary dispersion that were expelled and then repelled until taking form, experiencing gravity, and then releasing energy. The energy that escapes outwards is a lesser form (a hungry feeding) of antimatter that in time develops into black holes. This is another reason that the universe is continuously expanding (energy escaping the innerverse).

The radiant energy released by gamma blast that is not absorbed inward must be released outward. These are particles of subatomic and newly created atomic particles that are the ingredients (black holes) necessary to produce light. Away from the outer ring of the gamma belt is where the nursery of the stars was able to develop. Massive supernovas occur frequently. Black holes are at work creating galaxies of light and the process of evolution continues.

The energy released by the primeval germ expanding and putting into works an innerverse that was afterwards self-perpetual, was eighty to ninety percent of all the energy that existed, and a dosage or surge such as this was never again necessary.

The afterglow that we observe today began in this way. The art of self-creation and continuous evolutions was once again established. In this fashion, the stars are born and light is created. The universe of light finds existence outside of and away from the gamma belt and its inner void, establishing boundaries that allow these two parts to coexist and the innerverse to remain isolated.

THE WORKINGS OF THE INNERVERSE

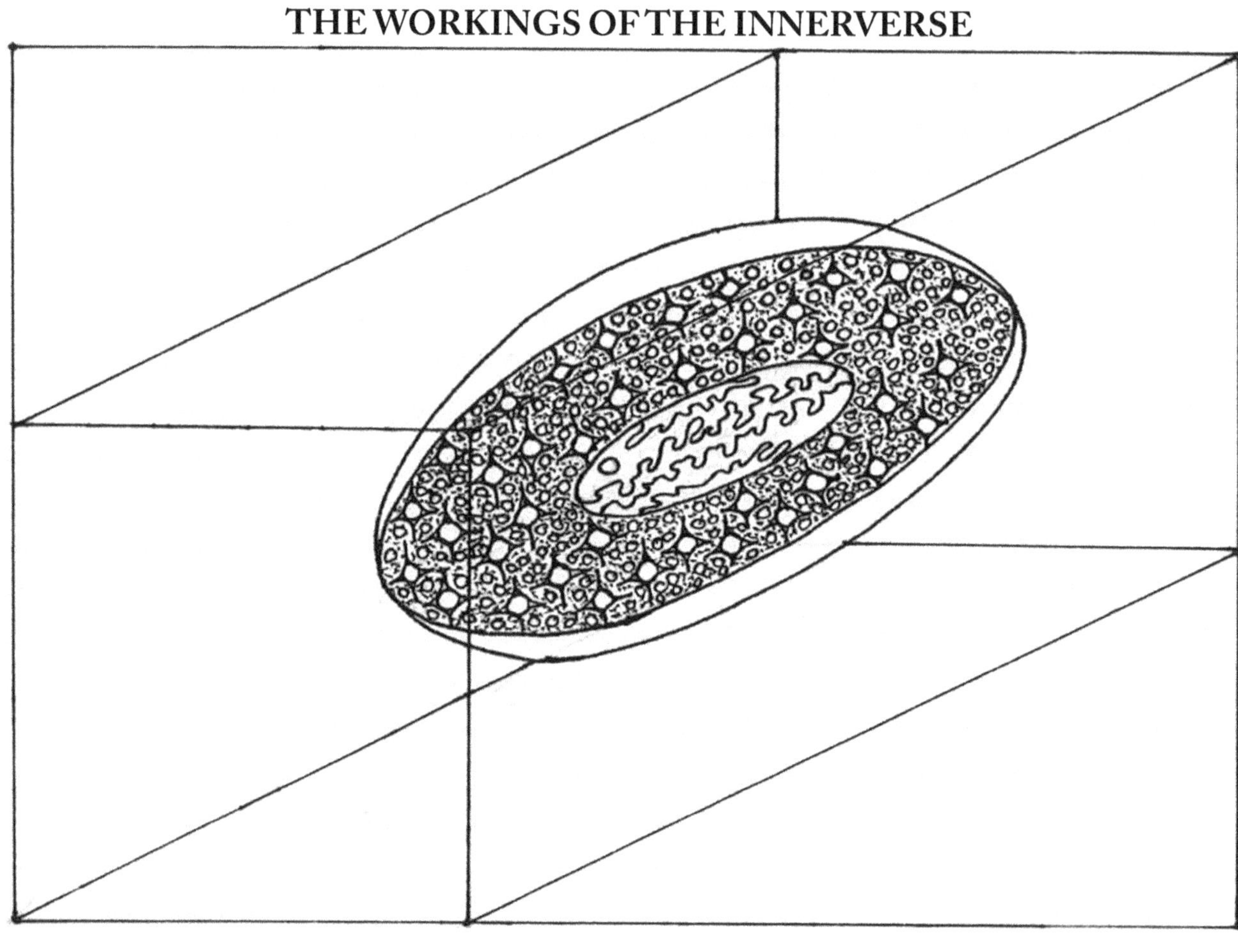

African Thoughts Concerning the Universe

The universe is a working mechanism that operates independent of humankind and because of this, its mode of operation is a great wonder. In this text we discover that there was an ancient African civilization that most definitely possessed knowledge of it, before the Greeks. Knowledge is a gift that is disseminated from RA, the star above, and the Greeks were ignorant of this fact. In their multiplicity of gods and goddesses this factor is not present. If it is then it is fanaticized within their mythology and the atomic value becomes lost.

Radiation and light, the two main forces of existence, are not factually mentioned. This intellectual void was passed on and is evident in the evolutions of the earth cultures that followed. It was like this until A.D. 1964, when two astronomers, Arno Penzias and Robert Wilson, in an attempt to detect microwaves from outer space, inadvertently discovered a noise of extraterrestrial origin. The noise did not seem to emanate from one location but instead it came from all directions at once. It became obvious that what they heard was radiation from the farthest reaches of the universe that had (possibly) been left over from the Big Bang.

This basic information concerning the reverence of light and its source is not found in the Hebrew concept of adoration or in any or most theological theories that are derived from the Age of Pisces. Therefore we can appreciate and accept that perhaps, two million years ago and more, Africans (hominids) began the quest of understanding the light. The depository of this comprehension was a nation called Egypt. Within their doctrines that relate to the light and a cosmic egg we discover that the basic signature of matter is the written symbol of RA. The circle and the dot that was visible to them in the observance of the eclipse is also the symbol of the atom.

There are different models for the structure of the atom. One of the first models was developed by a European. He proposed a model in which the electrons circle the nucleus in the same way as the planets circle the sun. Did this Danish physicist (Niels Bohr) finally come to realize that there is a generic connection between the sun (light) and the molecular structure (makeup) of the atom? Each orbit represents an energy level, which can be determined using equations generated by Planck and others. This model was later proven to be incorrect.

The accepted model is the quantum model. In the quantum model, we state that the electron cannot be found precisely, but we can predict the probability or likelihood of an electron's being at some location in the atom N: the principal quantum number, an integer value (1, 2, 3,..) that is used to describe the quantum level or shell in which an electron resides. The principal quantum number is the primary number used to determine the amount of energy in an atom. Using one of the first important equations in atomic structure (developed by Niels Bohr), we can calculate the amount of energy in an atom with an electron at some value of N The gap (Greek understanding) began to close as the order of the atom began to reveal itself. It was from this improbability concerning the placement of electrons that the idea of Xlectron and oblivion began to make sense to me. Today the circle and the dot are the accepted symbol of the (hydrogen) atom.

Within this universe that produces light is another, even greater force, radiation. For this knowledge to be found within the African (religious) doctrines of the light that are of prehistoric origin and not found within the logical concepts of the Greeks or in the Hebrew text (or those of the modern day) is indicative of the influence that the Great Ages have played in the universal awareness and in the development of humanity itself. These gaps and omissions have in their own way dehumanized us and we became less than what we were, unaware of who we really are.

The doctrines of separatism are "wrong," regardless of reason or religious belief. And to think that now these sentiments of separation have already been planted upon our moon! It is wrong for mankind to contaminate space with its mental viruses of racism or separatism. It must become clear that evolution and enlightenment are a part of a natural order that is also beyond the influence of mankind, and no one Earth culture or one race under the sun can claim dominance over all the others. This new age demands that the One must stand for the All.

The understanding of these two forces, light and radiation, has prompted the development of the schools of modern science. There are two (key) names that are associated with these forms of matter, Albert Einstein with light and Stephen Hawkins with radiation. Their understanding collectively opens the doors to understanding these fields in (numerical) formulas. I wonder now if the first formula of matter, in much simpler terms, is a repetition of those much more complicated. Shu is equal to Tefnut and Seb. It becomes the formula that reveals to us the order of existence: Sa=TS; T=Sa/S; S=Sa/T. Are we talking integers here? Was this the ancient formula for fission and fusion?

In the fashion of those that preceded them, these two giants of modern science sought to unravel the mystery of this universal order (unsuccessfully), and in their own way, they have laid the ground work for the conclusions that I have drawn within this book. The advantage that I possessed, and that they did not is that I am of African descent and they of European and so naturally they believed differently (cultural adaptation). By relating to my past I was able to understand parts of a cultural story that was of the same magnitude as the theories that dealt with light and radiation, and as we have seen this African version makes more sense than all the others.

To think that the greatest books and schools of thought were compiled and formatted without this knowledge is frightening. Begin with Isaac Newton, who is credited with discovering the laws of motion, such as: "a body at rest will tend to remain at rest until acted upon by an outside force." Was this really his discovery? Before Newton, the Greeks thought the same thing. De Lubicz notes in *Sacred Science* p. 42: "Besides, whatsoever an object may be in itself, it will either act of its own accord, or undergo the action of other things upon it." What we have here is the same concept, with one being a Greek (translation) understanding and the other that of an Englishman-and both are European. For too long they have been receiving the credit that is really due to others.

The conclusions (mockery) that they have established with their science and religions, which are obviously an act of (misunderstood) duplicity, confuse us. They copy from those they claim to despise and pass on unclear revelations of these "stolen" ideas. Cultures that are and were ignorant of the light are more often (racist) retarded and redundant, especially in their scientific findings. Light can never rest; it is always on the move, as is the search for truth.

Perhaps the most famous light formula of the last one hundred years is E=MC squared. It was the brainstorm of Einstein, who realized that light can be valued in terms of threes with each of the three possessing numerical value. It was as though he looked up at the sun and had this revelation. However, the sun was not the source of his inspiration and although his conclusions were worldly acclaimed he was unable to completely understand the origin of the universe. He developed the most astounding theory concerning the speed of light, never completely understanding its origin.

By combining the ancient Egyptian text and the findings of Stephen Hawkins concerning dark radiation and black holes, it was clear that matter and radiation could be destroyed. Hawkins had demonstrated (to me) that light and radiation are subjected to another, more powerful force and that they also can be consumed. If this force is oblivion, then the ending was ordained. I at once realized that this ending was also the source of beginnings. Next I connected the creator of the light to a black hole and it was evident that this is where the light originates. I came to this conclusion by applying the laws of dualism.

Now I had a full understanding of the origin of and the demise of the light. The mode of operation was much clearer in my mind. If I could figure out the source of radiation, I would have the beginning. Again I related to the ancient Egyptian text that speaks of a source that produced the light. This source had to be radiation, aka the cosmic egg. I was in pursuit of a force (radiation) that was also self-created, but from what? (Would I be able to explain this theory-"Rasaurian Star Theorum-to Einstein?)

I was able to dismiss the Big Bang theory because it did not involve true origins or include concepts such as oblivion. Besides, it was obvious that people had been trying for years, to develop a working model of the beginning of the universe and the Big Bang had been their inspiration. I updated everything that I understood about cosmology and twenty-five years after I had begun, it came to me.

Using every idea known to me concerning the light and all the new findings of astronomy and space exploration, I put together this (Rasaurian Star) theory from the hundreds of clues. At its very first annunciation it was said that the cosmic background radiation is equal to two-thirds of the available space in the universe. This meant to me that the universe of light must exist in the other remaining third. The idea of an innerverse and an outerverse began to seem feasible. That the universe is home to both, shed new light on this old mystery. What if the universe is an integer, consisting of oblivion, an innerverse, and the outerverse that make one whole universe?

Now I was armed with a formula (information) that really worked. If I could understand any two parts of this formula, they would then reveal the other parts. I knew of the light and radiation and black holes and of oblivion when I applied the concept of self-creation as in the ancient Egyptian text, voila! Everything had already been worked out for me, leaving only the application of science and the understanding of religion and how they must be one. It all began to make perfect sense.

Plato taught the Greeks that "Life is, was, and always will be." This he learned while in Egypt. I thought that if this is so, then life (energy) must always be present somewhere. By applying life to oblivion, radiation to the innerverse, and light to the outerverse, the universe of heat and movement began to make sense. Under normal circumstances matter cannot be destroyed. There is a line in the ancient Egyptian creation text: When a star ("ceases to exist it has left its station") is captured by the gravitational pull of a black hole its energy will be consumed (likewise with the black hole), it is collected in Oblivion.

Now, what if all the light and radiation that exist were collected in Oblivion-would that be the end of it all? That is, if all the energy that is existing were to be engulfed by the gravitational pull of a super black hole and then in turn that super black hole is subjected to the laws of extinction as explained by Dr. Stephen Hawkins; where did the energy go? African cultures believe in the repetition of life, which must always find a way to exist.

This concept now began to have terrific value and significance because I understood that life would jump-start itself all over again. I needed only to figure out how this could happen, recalling the thoughts of Ausaures: "In the whole of the Universe nothing is ever added and nothing is ever lost." Meaning to me that everything must be accountable at all times. That means every molecule or particle that has ever existed must be ledgers in this theory.

After reaching several dead ends I realized that within the magic of the atom was a particle (proton, neutron, electron) that could serve the purpose of jump-starting life anew, the electron. In quantum mechanics the electron "cannot be found precisely" and it is thought that electrons have the ability to move interdimensionally. Armed with these facts I needed only to imagine the demise of matter inside a black hole. I pictured the atom being decimated until only the electrons survived as harmonic strings or strands of energy and then they also are collected into oblivion. It is known that Albert Einstein, was unable to appreciate the findings of this "string theory." Here, however, I found practical use.

Xlectron is the strand of energy (the results of atomic decimation) that (has the ability to) escapes from oblivion and begins anew an order of life that is based on heat and movement. Because light and radiation are mirrored images of each other, the order of self-creation is similar (ancient Egyptian text). Having escaped oblivion, Xlectrons very first experience is weightlessness. The void of black space had not yet come into being. At this time, there was only Xlectron. it existed as a string of fluctuating energy with a neutral charge. Uncomfortable with this feeling of weightlessness, its spirit was aroused. It experienced intrafarance and began to change. It sought out itself, took form (egg shape), and began to emit waves of repulsion allowing it to feel itself and overcome the weightlessness. This was a magnetic function and the laws of dualism were set into motion.

The void of dark space develops hereafter, in proportion to the growth of this newly self-created life form. Repulsion, the emissions of energy which has a magnetic function (the results of attraction) becomes the initial force that extends the dark void. These emissions propel outward, expanding the universe. Xlectron had now become a living cell or being. It became the primeval germ.

It needed to feed and so it expanded itself in an explosion that cast away the shell that had once contained it. This was the greatest release of power ever produced in this newly forming universe. The fragments of shell were pure antimatter, directly from the eternal source. They traveled outward until reaching the end of the flux of magnetic repulsion and there, they orbited and experienced gravity, collapsing (upon themselves,) and produced gamma explosions the energy of which was released inward and absorbed by the primeval germ. This was the completion of the innerverse that allowed the development of subatomic space as particles of nourishment for the continuous growth of the primeval germ. This is the germ of radiation and from its residue comes the light.

From its inception the innerverse has occupied two thirds of the available space. Its outer barrier is the gamma ring or belt that surrounds the germ of life. Small portions of antimatter and energy emissions escape the inner attraction and continue onward into the remaining one-third of the universe. This one-third or outerverse is the home of atomic matter and the domain of the light. The nurseries that produced the galaxies of light are found outward of the gamma ring and the light is not known to exist within the innerverse.

The very small bits of antimatter that survive beyond the gamma ring become black holes that also have the gravity to collect subatomic particles. Upon their combining together, light is produced. This was the scientific understanding of the ancient Egyptians. And their (religious) belief in a star (RA) was how it was disseminated to its followers.

THE REALIZATION OF THE INNERVERSE

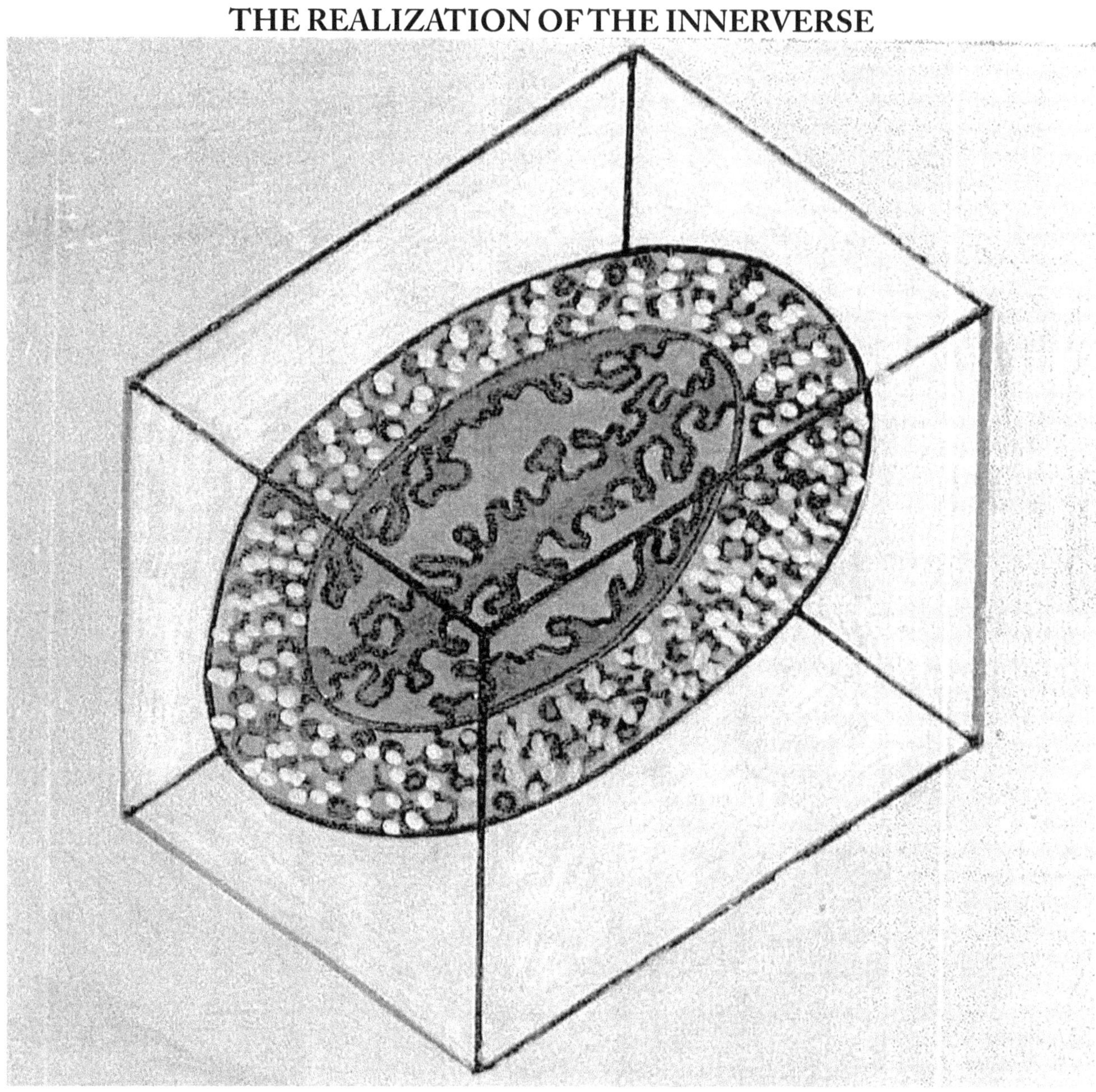

Chapter 15
The Origin of the Universe (The Innerverse)

All the major forces of the universe are of one origin. Necessity becomes the order by which these forces reveal themselves. Existence demands that energy be spent, existence is realized by the expansion of energy that must have a source. In the case of our universe, life is always associated with heat and movement. The energy that drives the entire universe is the result of the self-creation, of radiation that produces magnetic attraction and repulsion along with heat and movement.

In the beginning there was energy; its name is oblivion. And in the end there is energy; its name is also oblivion. Life is derived from power and power is the result of energy (E=PL). The first law of energy is (always) to exist and expand. Xlectron, after escaping oblivion, became a source of power. Because this power was dissipating, form or shape was necessary for containment and the feeling of weightlessness (then) must be overcome. By the emissions of controlled waves and rays of magnetic subatomic particles in all directions after taking shape, these two basic needs were met. This is the origin of subatomic space and the story of the self-creation of the primeval germ.

Energy that was lost and now being spent was in need of replacement. This created hunger and the need for replenishment. All the energy was being dispensed outward, allowing stabilization and growth within the newly formed egg until it was forced to expand beyond itself. This created a great explosion that hurled shell bits in all directions, expanding the universe as they traveled, until they finally reached the pause of magnetic repulsion. There they experienced gravity and bits began to collapse in on themselves and create explosions of gamma force. A ring (of gamma orbs) now surrounded the germ of life. It allowed the recycling of energy inward and the replenishing of itself. The innerverse was in this fashion created and the laws of motion were set into being.

Primary forces include source, repulsion and attraction. In this way, magnetic activity establishes the laws of attraction and repulsion as the oldest of laws concerning heat and movement. These laws allowed for life (energy that is contained and controlled) before the establishment of gravity and atomic substances and are developed in accordance with the appearance of and growth of radiation.

What is Radiation (C.B.R.)?

Radiation is an energy form that is the result of supernatural occurrences on, a universal metamorphosis involving beginnings and endings. It is the molecular appearance from the source of all (oblivion) creation. It is the energy that expands the universe outward and uniformly. It is a living, growing force. Because of this, it was thought to be a germ, the germ of life, the primeval germ of existence.

Chapter 16
The Big Bang
The Big Bang of Radiation

Existence requires the spending of Energy. In the fashion of self-creation this energy is realized through the completion of the gamma belt or ring. This process is begun by the germination that is occurring within the cosmic egg. Evolution is about to introduce itself with an explosion from an expanding and exploding egg. The shell is fragmented and is discarded violently outwards: The germ of life has increased in size and "the void of cold dark black space is expanded." The antimatter released in shell fragments travels until reaching a barrier of "pause" (stop) and gravity is experienced, causing these fragments to collapse upon themselves and explode, creating gamma explosions that fills the inner space with subatomic debris that is consumable and replenishing.

This has created a space germ that has (heat and movement) a nucleus surrounded by an exterior of orbiting orbs of gamma antimatter. It is self-containing and it and the barrier consist of two-thirds of the known universe. Thus the egg (surrounded by the gamma ring) becomes the symbolic depiction of the innerverse. In this story the forces of the universe reveal themselves with clarity and purpose, and definition is possible.

THE REALIZATION O FTHE INNER AND OUTER VERSE

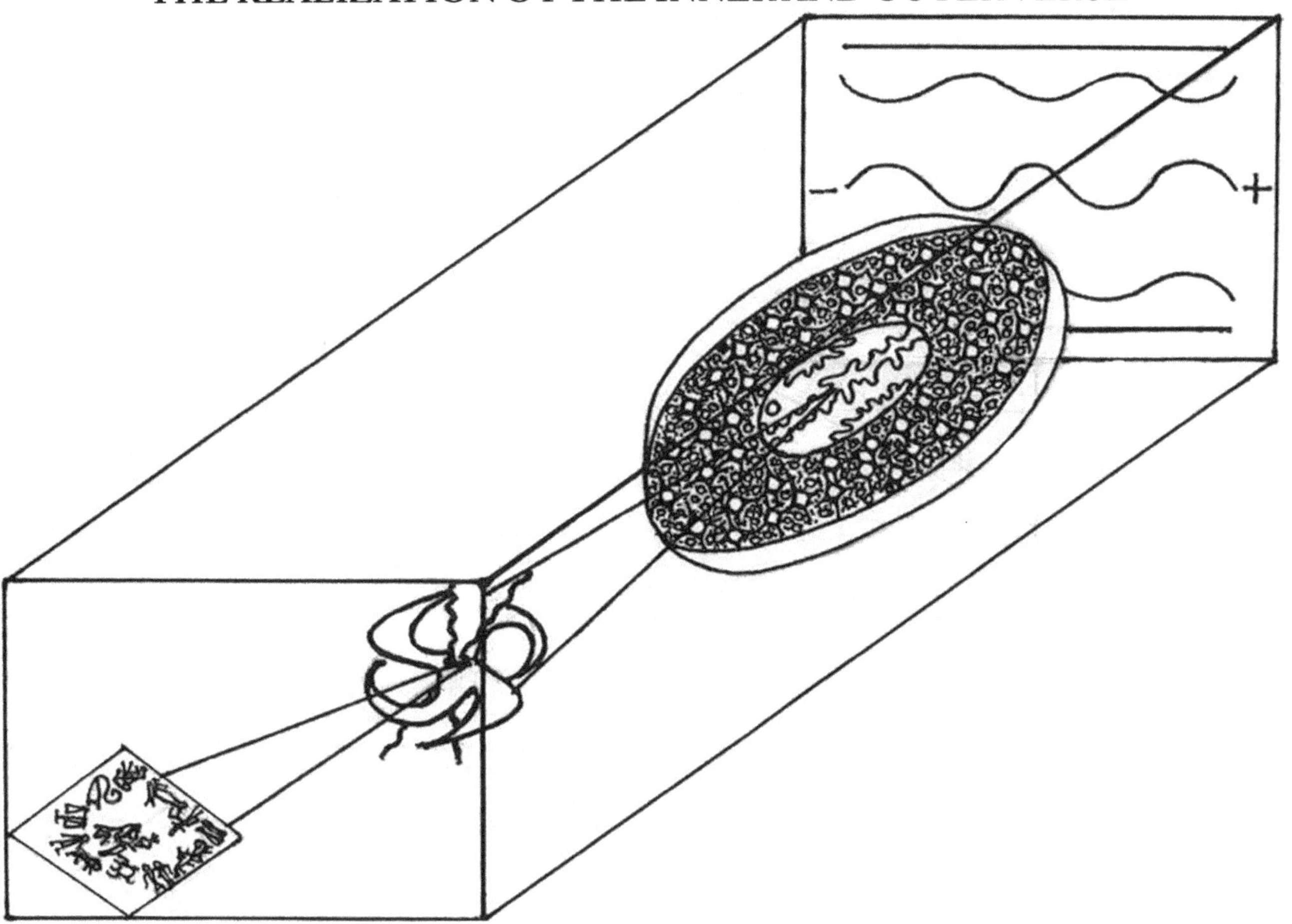

The Big Bang of Light
(Outerverse)

At the very edge of the outer gamma ring lies the baryon plasma barrier. Escaping particles both subatomic and atomic began to form within the remaining one third of the universe. It is here within the nursery of stars, where millions and millions of black holes exist, that we have a continuation of Big Bangs where production of the light takes place. It takes place here within the outerverse. If I am allowed to make one adjustment to this theory, which is that light is the result of It and that "It," is still continuously occurring, then this Big Bang theory makes perfect sense. If the Big Bang is the story of the beginning of radiation and Light, then this is how and why it seems to have occurred.

Let us quickly review this theory. The Big Bang is the dominant scientific theory about the origin of the universe. According to the Big Bang theory, the universe was created sometime between ten billion and twenty billion years ago from a cosmic explosion that hurled matter in all directions. In 1927, the Belgian priest George Lemaitre was the first to propose that the universe began with the explosion of a primeval atom. His proposal came after astronomers observed the red shift in distant nebulas to a model of the universe based on the theory of "relativity." Years later, Edwin Hubble found experimental evidence to help justify Lemaitre's theory. He found that distant galaxies in every direction are going away from us with speeds proportional to their distance.

The old Big Bang theory (with all of its short comings) was initially suggested because it explains why distant galaxies are traveling away from us at great speeds. The theory also predicts the existence of the cosmic background radiation (the afterglow or the leftover from the explosion itself). The Big Bang theory received its strongest confirmation when this radiation was rediscovered in 1964, and most recently with the findings of (C.O.B.E.). Again New Age application demands the developing of theory that is based on insight of the past and updated by modern technology. The more the facts support a theory, the better the chance of it's being true.

When C.O.B.E. looked back into the past it found the subatomic particles that first fluctuated and would evolve and then eventually become the baryon plasma barrier, the source of the hydrogen atom. This barrier's placement (the outer ring of the gamma belt) strongly suggests that the atom can exist only within the outerverse and these atoms identify themselves as stars or other sources for the light. All the matter (that exists) within the outerverse is the result of energy that escapes the innerverse and together they make up the entire universe. The source of the light is revealed as radiation. The source of radiation or beginnings and endings is oblivion. In this theory of accountability, "nothing is ever added and nothing is ever lost."

THE UNIVERSE WITH STARS

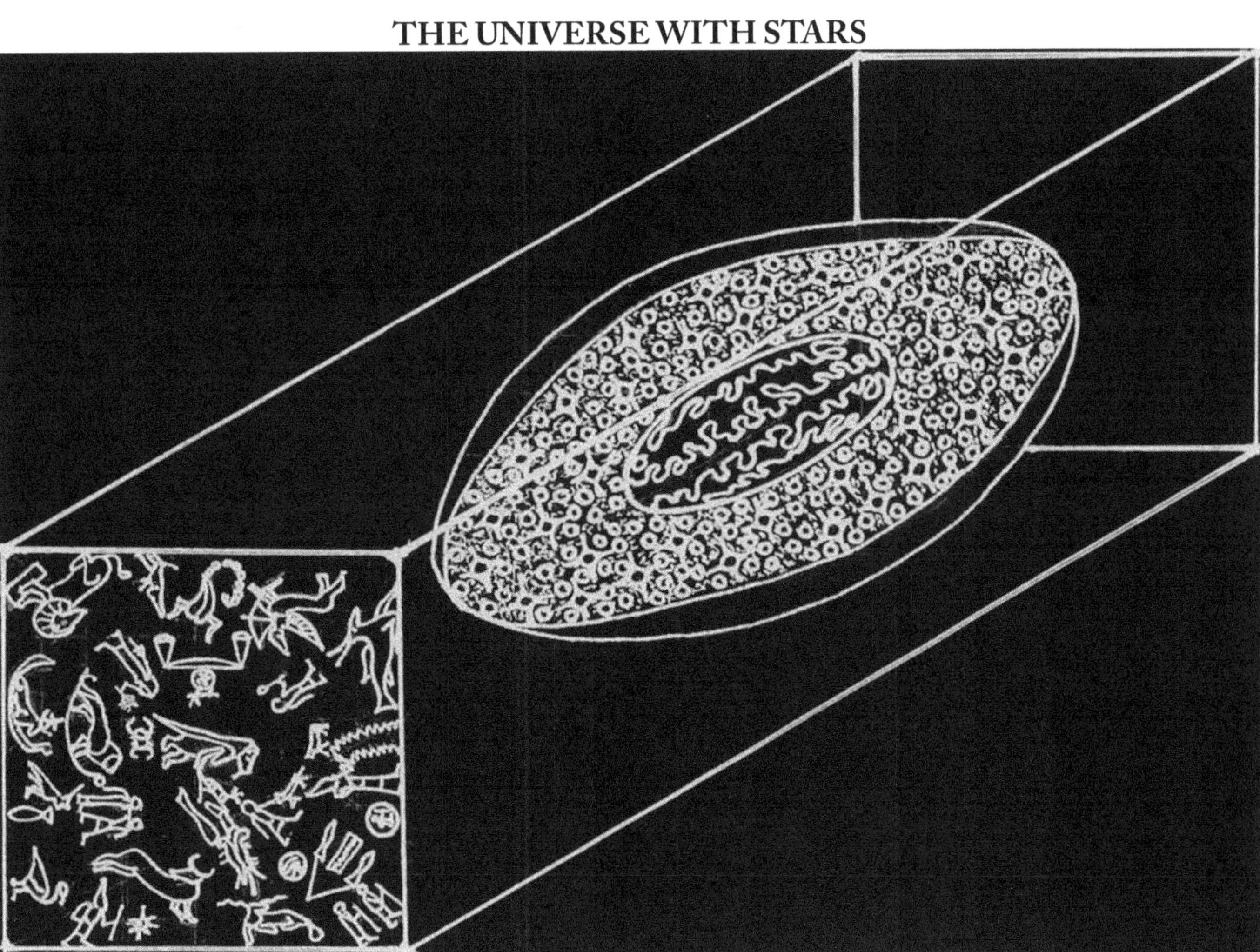

Chapter 17
The Laws of Magnetism-Expansion, Attraction, and Repulsion—and of the Void within a Void and of Being

In considering the concepts of the magnetism, which produces the anomalies of expansion, attraction, and repulsion, five phenomena must be understood. They are the (two) voids, heat, movement, and being. Without any single phenomenon the others would not function. They are interdependent upon one another and their presence allows purpose, function, order, and law.

One void is the result of being, and being is the product of the other void, which is oblivion. That is, the darkness of cold space is a necessity of being, as are heat and movement, while the void of oblivion is the origin of being. The laws of magnetic expansion, attraction, and repulsion begin with the understanding of these five Anomalies.

Something" is always the result of something else, or something (change) is always the result of the combination of these five anomalies, (intrafarance). Change is the one constant within the universe. It occurs because of the interaction of these principles.

Oblivion is the void of endings and beginnings. Inside a black hole where matter can be destroyed, a door opens to oblivion. The atoms are destroyed there. They are disassembled and decimated and their atomic energy is absorbed. This place, oblivion, is the capacitor of existence.

Capacitance is the ability to collect and store energy. This energy that is collected can also be released or discharged. From this energy that is collected and discharged come the laws of attraction and repulsion. Harmonic strands or strings represent the quantum result of the total decimation of the atom, leaving invisible electron lines of magnetic flux.

Opposition or Repulsion allows the expulsions of this energy in accordance with the laws of universal capacitance. Which are mirrored in the realization that even black holes can be absorbed into Oblivion by the process of "over consuming." Why would this not also apply to Oblivion? When the void is completely filled to "capacitance" energy must be released.

"Xlectron" is this energy in principle; that is able to release it-self and then seek out it-self and experience the phenomenon of "Being" or energy that has the ability to self-create. This would appear to be beginning of radiation linking "it" to the appearance of being. The "red" afterglow that was recently confirmed by C.O.B.E. suggest, that one, is the result of the other.

"The super telescopes have spotted the beginning of the Universe and most, scientist and researchers having knowledge of the most recent, findings concur that the birthing cradle of energy starts here. (Cosmic Background Radiation) These five players represent the concept of unlimited power turned on."

The cosmic background radiation is the equivalent of the power that was turned on in the very beginning, similar to a light being turned on, never turned off, and able to burn indefinitely. Then we must add the idea of this light having the (ability) power of perpetual burning (growth) and of course, intelligence. Because of C.O.B.E. this anomaly is now identified as the cosmic background radiation.

Radiation is the beginning of everything, this side of oblivion. It is here that the laws of attraction and repulsion begin. Motion is a requirement in the separating of the collective energy contained in a motionless cold void, such as oblivion.

In order for this parting to occur, change, the ability to alter and rearrange, must be present. This is accomplished by the inclusion of the final two players, heat and movement, which allows the development of form and molecular structure, radiation.

Magnetic induction (form) is then induced. Radiation is created, and both are the results of being or self-creation. The universe begins in this fashion. Intrafarance is realized and the inner-void of black subatomic space begins. Laws become necessary in the establishment of uniform expansion, which is the result of the heat and movement that is emanated by this force of cosmic radiance.

Magnetic attraction allows Xlectron to seek itself out and magnetic repulsion allows for the overcoming of the sensation of weightlessness. These two fundamental needs are the first steps taken in understanding the laws of heat and movement or the concept of being.

The "trinity" of radiation or the cosmic background radiation are heat, movement, and being. The instruments that C.O.B.E. carried aboard confirmed the heat and movement. However, they were unable to detect being. These billion-dollar instruments could not be programmed to perform such a task.

What they did do is pinpoint an anomaly that clearly represents the beginnings of radiation. C.O.B.E. and its instruments concur that this is the earliest and the oldest anomaly within the universe.

Before there was light or stars there was cosmic radiation. It occupies a portion of space in which the light is absent. In its need to perpetuate itself the laws of motion were implemented. Having escaped oblivion and overcome weightlessness, and upon taking form and becoming a being, the need for replenishment became a necessity.

To sustain it-self, it expanded with a self-induced explosion in which parts of itself were blown outwards, creating a ring or belt of energy of antimatter that allowed it to feed of itself. This great explosion that occurred then was the largest discharge of energy the universe has ever experienced. (The birth of dark matter and anti matter)

Its results were the natural expansion of the void and the placing of the gamma belt or gamma ring that enveloped it. This envelopment is proportional to the growth of the anomaly and is the realization of the innerverse or space that has the absence of light.

These parts of discarded energy are bits of pure living antimatter that upon reaching the barrier of magnetic flux or pause, experienced gravity and collapsed upon themselves, creating nuclear explosions of gamma blast.

The introduction of gravity and the gamma barrier established the innerverse and the outerverse. Gravity is the result of antimatter collapsing upon itself and producing gamma explosions that released nuclear energy inwards and outwards.

The energy that was released inwards was (is) absorbed by the self-created being or cosmic radiation. The energy that escapes outwards, beyond to the photo baryon plasma barrier, as bits of antimatter that possess gravity eventually become black holes.

Stars and light are the results of black holes. Until the collection of all the mentioned energy (once again?) into oblivion, these forces will propel, expand, and adjust accordingly, the continuously developing and expanding universe.

1 The circle: light would travel in waves of repetition and time would continuously remain unchanging. If the energy in light waves is not allowed to disperse, it accumulates.

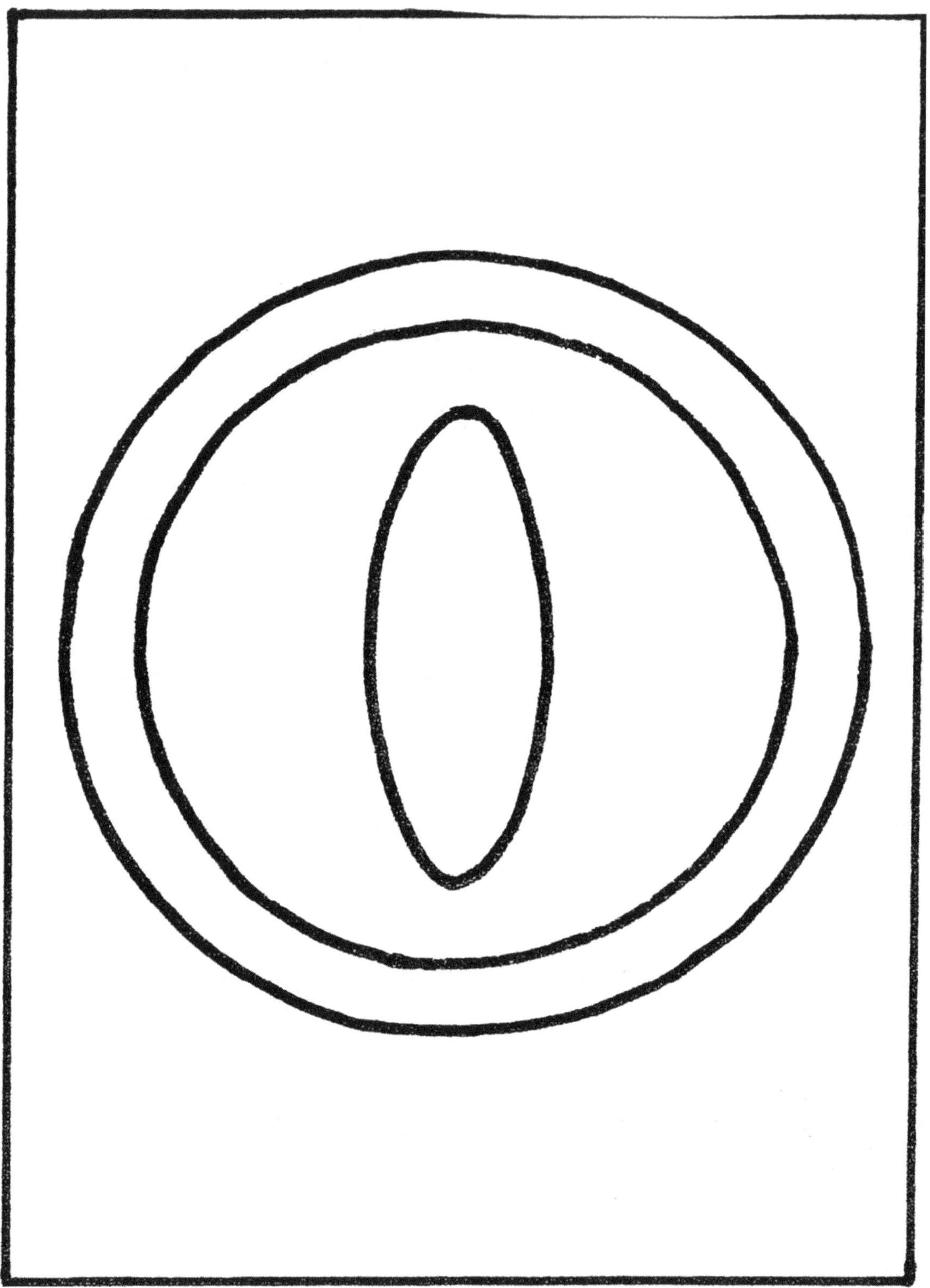

2   The triangle: light that is contained in this form would be subjected to three ends and open three doors of disruption.

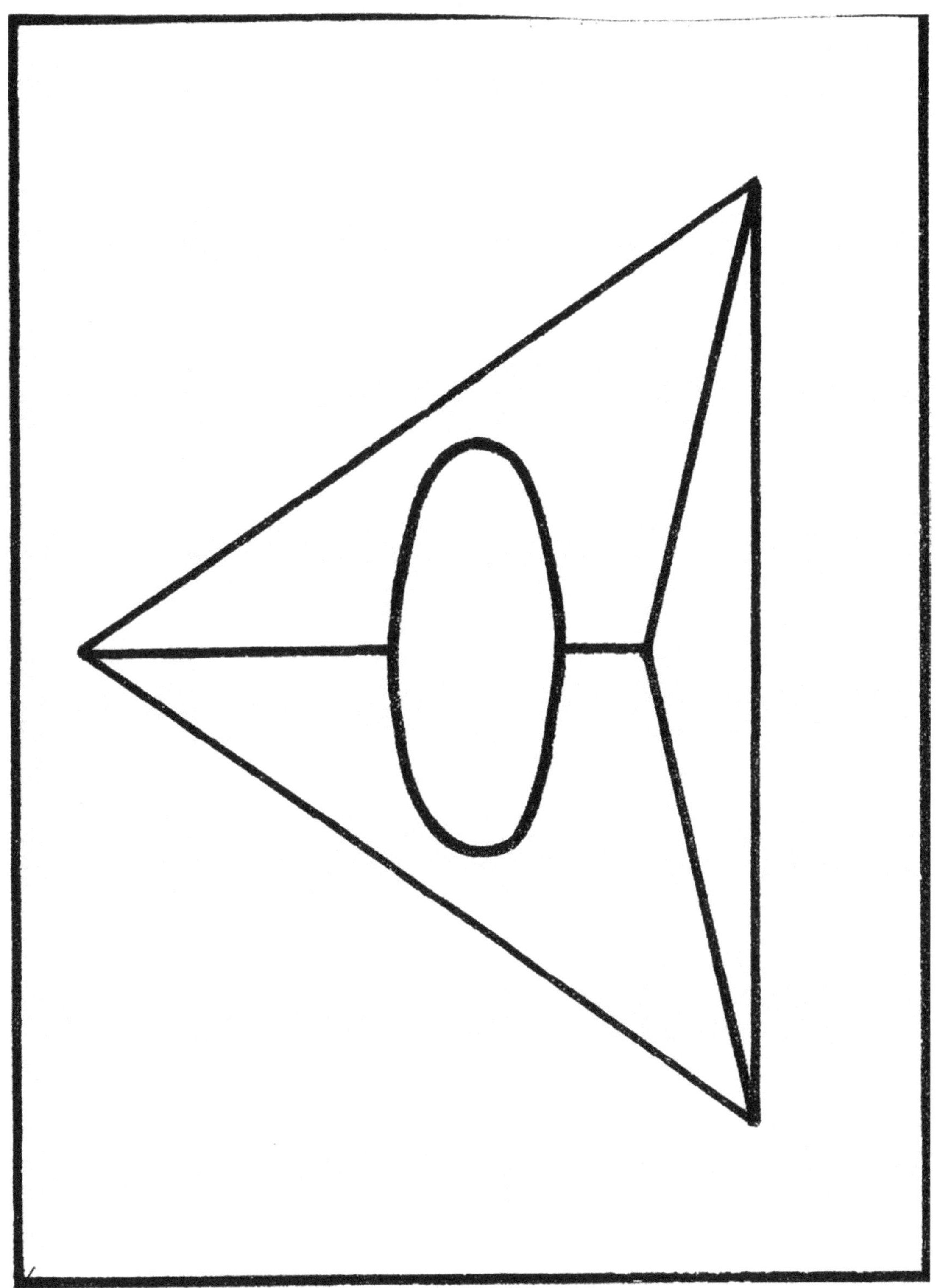

3 The square: in an ever expanding universe the square container allows light the possibility of endless travel. Time would be infinite. Where light can travel unimpeded in straight lines in accordance with the growth of the primeval germ.

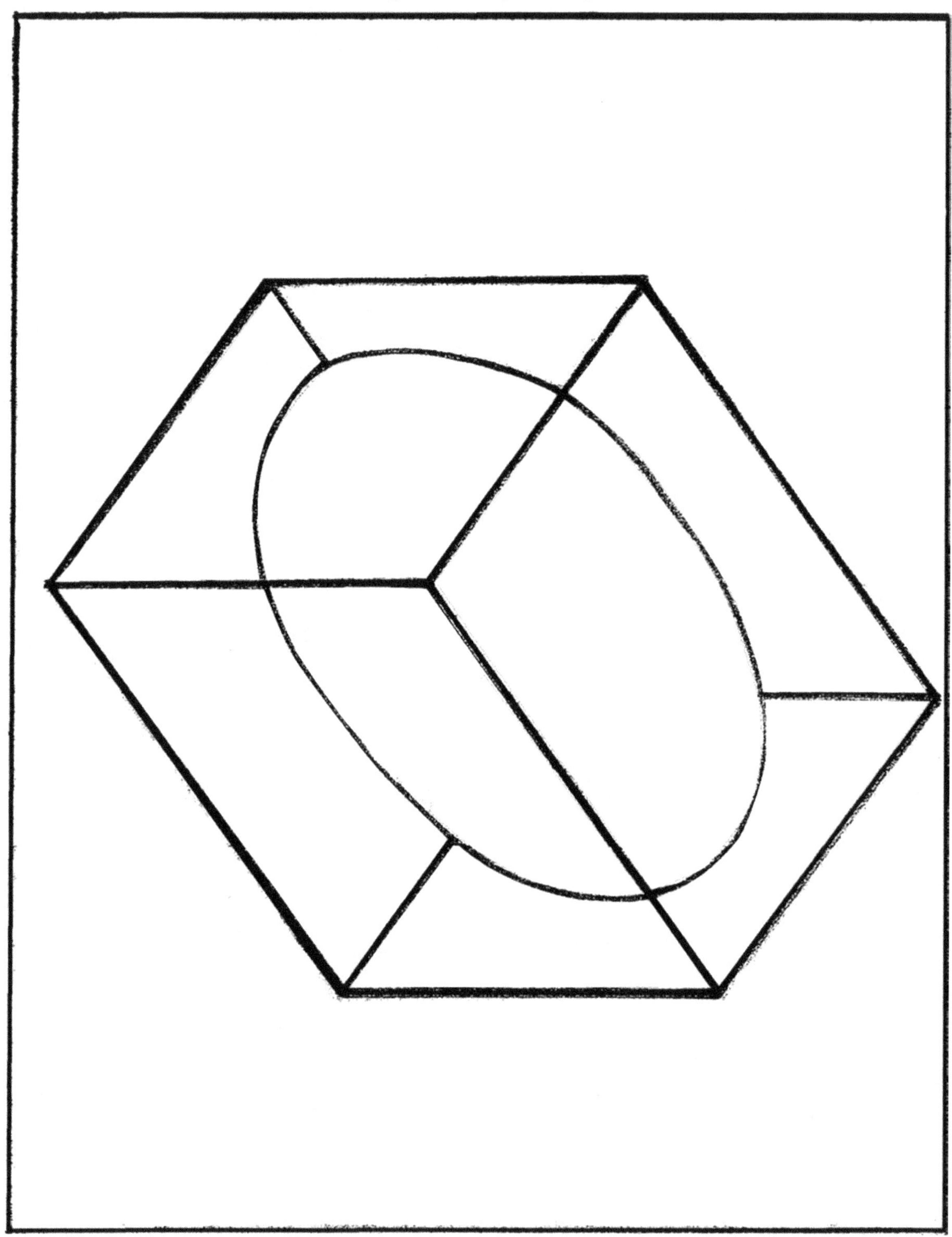

Chapter 18
The Order of the Innerverse and the Outerverse

Oblivion: the place of collection.

1: The spark or harmonic string is freed.

2: It seeks out itself. It takes form (an egg) and begins to self-create.

3: It cannot feel itself; it is weightless. Emissions of subatomic magnetic repulsion particles. The weightless one can feel itself and becomes a living thing (germ).

4: The depths of the universe take form. The darkness of space is realized.

5: It evolves and evolution has begun.

6: The order of self-creation is established. The discarded shell is fragments of antimatter.

7: This antimatter experiences intrafarance and the gamma belt is developed into orbs of pure energy. These orbs produce gamma blast that releases energy that sustains perpetual growth.

8: The cycle of continuous evolutions is created along with the innerverse!

9: The gamma orbs develop gravity (the force that ignited them).

10: Gamma energy that escapes outward as even smaller bits of antimatter create the seeds that begin the outerverse, and the nursery of the stars is developed.

11: The black hole is at the center of every known galaxy. It is believed to be one half of one percent of the entire mass. Light is produced there.

12: The black hole is a door that opens and absorbs light. In its turn it is absorbed. and returns to the place of collection.

13: Oblivion.

The ancient ones believed that life finds and always will find a way to exist.
The end is the signal of a new beginning.

Chapter 19
The Order Of Particle Existence

Strings or flat electron	negative charge
Orb or disk	proton neutron
Orbs or dualism	proton= +, neutron= -
Compound or grouping	natural attraction, natural repulsion
Intrafarance or the changing	fusion/fission
Neutron orb with a plus and minus electron string	intrafarance
Proton orb with a plus and minus electron string	intrafarance
The atom is produced	intrafarance

Chapter 20
Electrons

Electrons are the staple of heat and movement.

If the electron moves inward of its orbit, the atom will radiate.

If the electron moves outwards, the atom absorbs.

All particles and molecules have electrons.

Electrons are unstable and unpredictable.

Electrons are capable of dimensional movement.

They have the ability to appear and then disappear randomly.

Electrons are the only particles or molecules that can exist in subatomic or in atomic states and also within the void of oblivion.

Chapter 21
The Cosmic Egg: Time in Terms of Existence

Time and the outerverse!

Time: begins when the creator of light takes form and commands the power.

Time: is an interval in measuring the speed of light as it travel, from its source to its destination.

Time: is composed of atomic particles of radiant energy that formed outwards of the gamma belt.

Time: is nonexisting within the gamma belt that surrounds the innerverse.

Time: is the witness to the orderly expansion of the universe.

Time: is ethereal matter made visible by the heat and movement of photons.

Time: ends, when all the light in the universe is absorbed and no longer visible.

Forever: the life span of a star, usually billions and billions of years.

Forever: is not the end of time.

Forever: as long as our sun burns and emits energy.

Forever: is eight to ten minutes away the distance in light time to the sun from the earth

Forever: is a message from a star that is 271,000,000,000,000, miles away. At time of its light's arrival on the earth, the star is extinct.

Forever: is a signal of light that complements time.

Always: the never-ending cycle of germination. As it occurs within the innerverse it is older than time and perpetual.

Always: is not applicable to light matter or radiation in the normal sense; it is applicable only to the term oblivion.

THE CONTINUATION OF THE DEMISE OF THE CUBE OF EXISTENCE

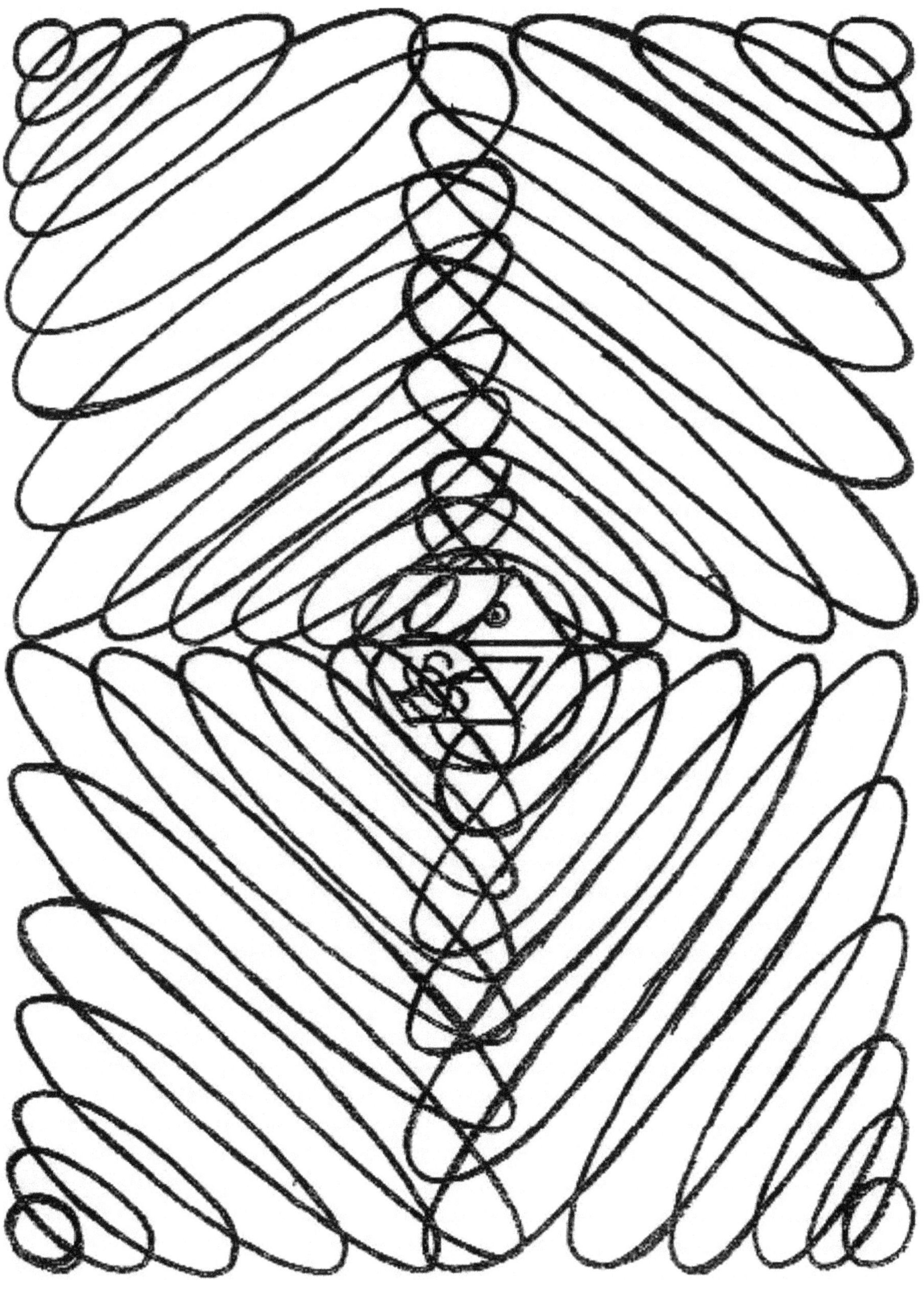

ENGULFED BY THE RINGS OF DELLERIUM

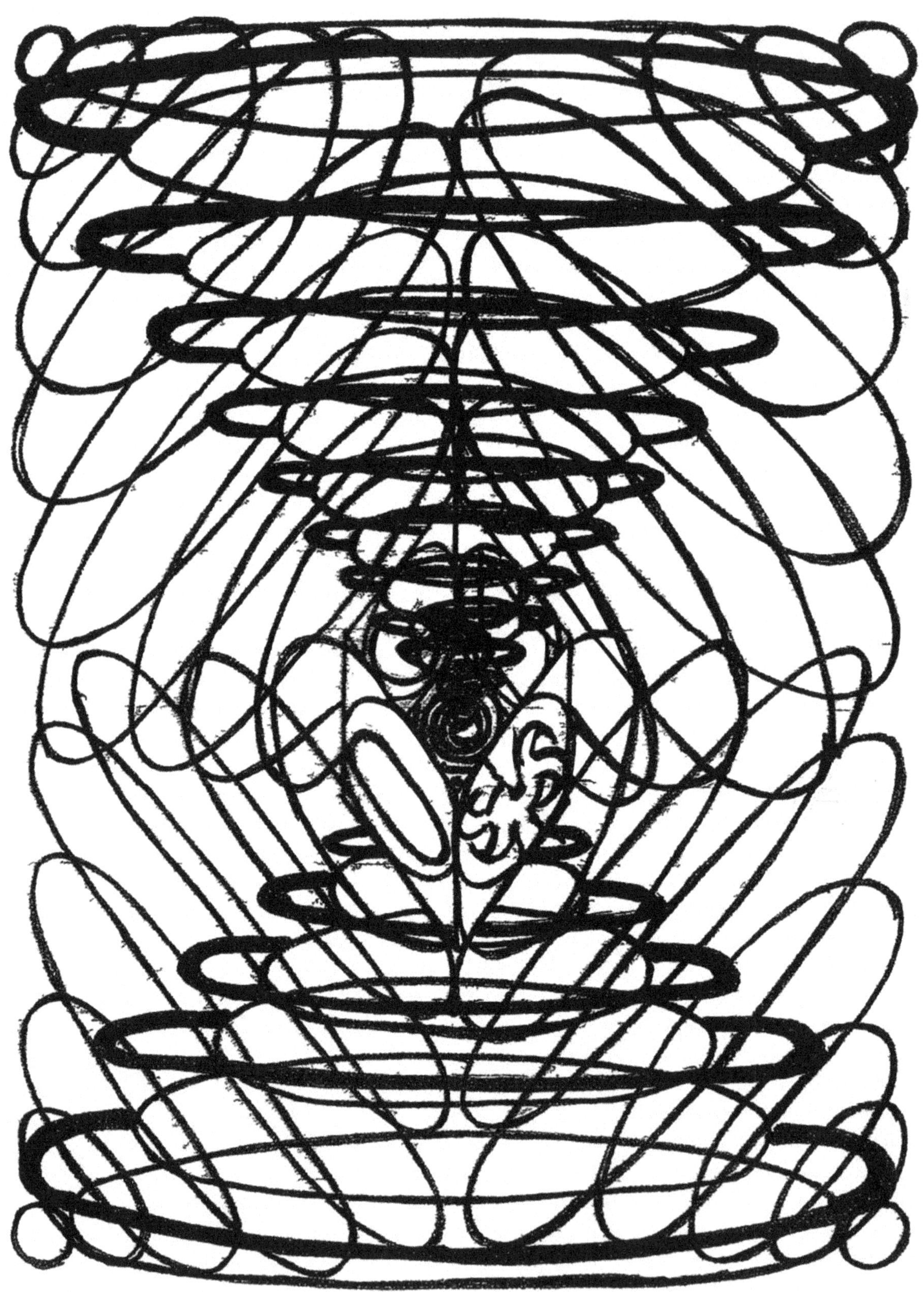

DECIMATION AND ABSORBTION

THE FINAL CHANGE/REARRANGEMENT OR ALTERING

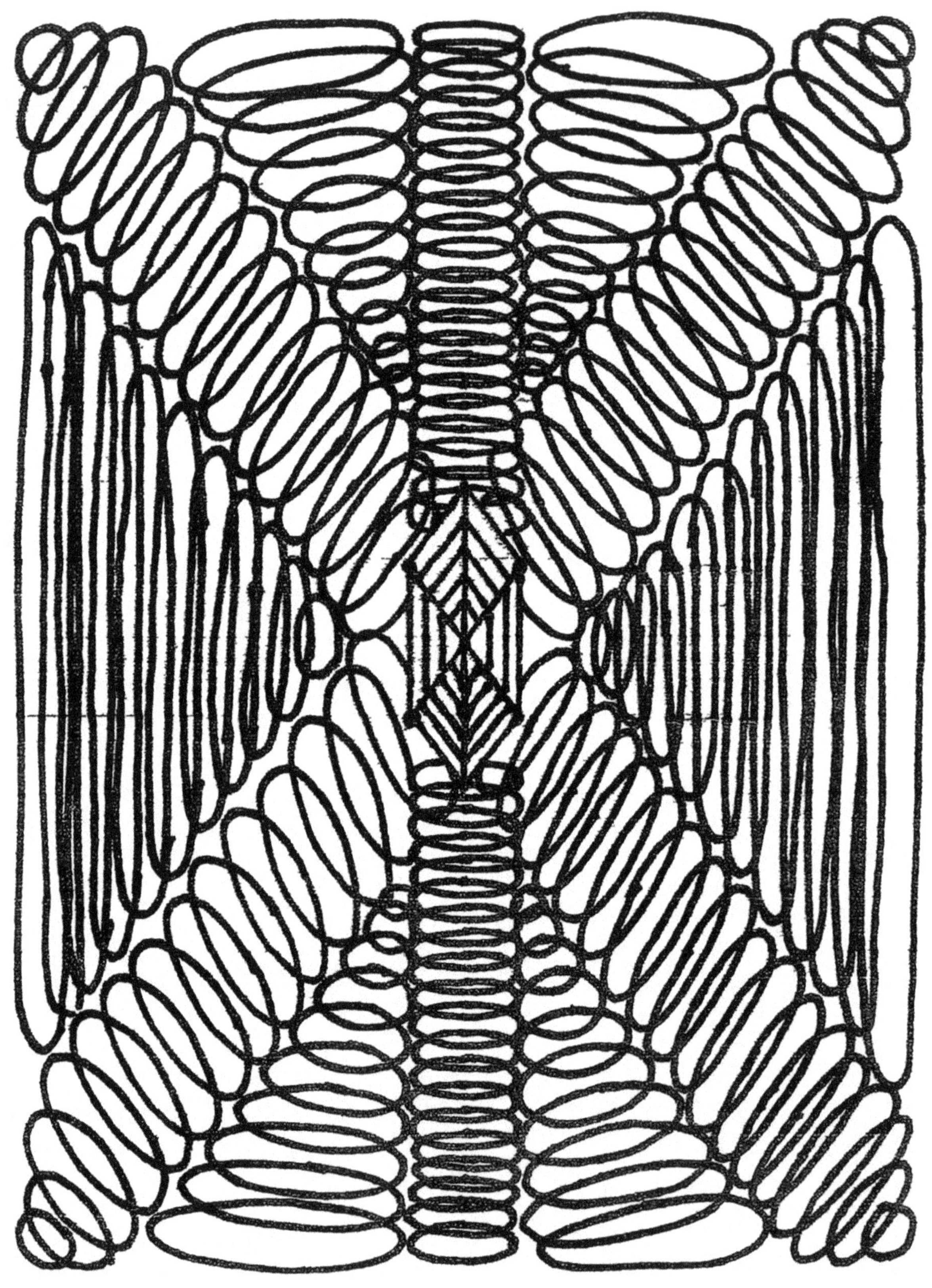

OBLIVION IS ACHIEVED

Chapter 22
The Cosmic Order of Numbers

O: The number of the all or oblivion.

1:      The all being represented as one or one part of the all. One indicates many. "I was alone at this time; there were no others with me then."

2:      The all reflecting the dualism of being. The extension of the source in opposite directions. A state of being that allows existence in two places at one time.

3:      The all representing itself as a trinity of forms. The triangle. The three parts of a circle: radius, diameter, and circumference. The joining of three to make the one.

4:      The all "squared"; uniform growth and precession. Indicates that more is to come.

5:      The all expressing uneven development. The number of accumulation or half.

6:      The all is now a duplication of itself. As above, so it is below. The "all of alls" or the mystery of it all.

7:      The all reveals itself. The number of the light.

8:      The all is cubed. The number of Infinity.

9:      The all is now completed in this number. The mystery is revealed.

10:      The all is mirrored. It is the number that is symbolic of the inner and the outer universe. Five of the ten are Radiation and the other five are Light.

Part 3

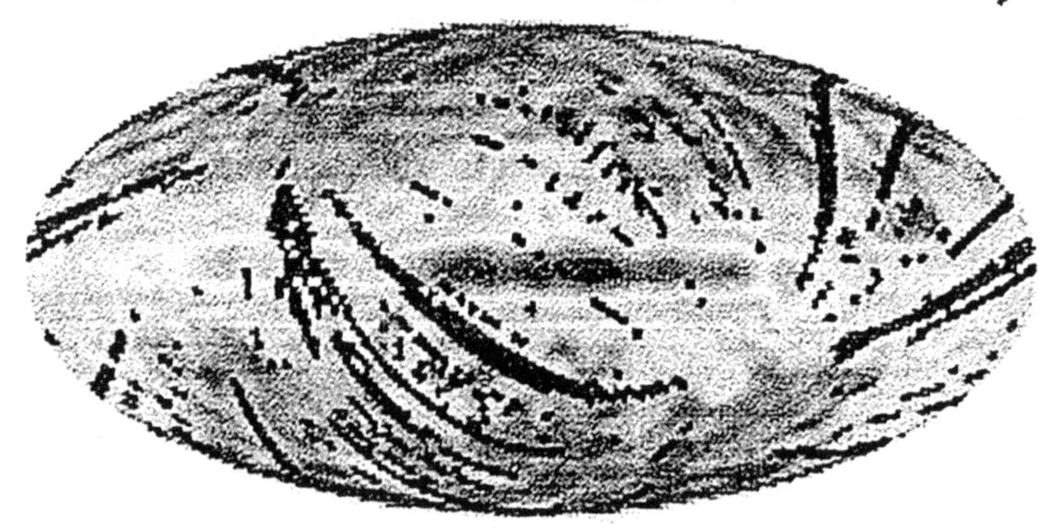

"THE ONE-THE ALL"

Chapter 23
THE ONE-THE ALL

Oblivion is a storage cell (in the saga) of life forms that are the result of heat and movement. Its existence is relevant to the laws of dualism. This was the home of Xlectron when the spirit of intrafarance combined with it. It was able to free itself from the void of Oblivion. It was and is the all of alls in one being. The creator of the light says: "I was alone at this time." Xlectron was also alone, but it carried the germs of all the others that had been collected within oblivion. When it combined with the spirit that never dies (intrafarance) it became "God" born anew.

This is the beginnings of being, or the rebeginnings of being. It was personal, and it was collective. It is how All can become One. This is meaning of the All of Alls, the story of everything that exists, from oblivion to oblivion. It is a tale of billions and billions of years. To mankind, the earth (Seb) is the All and its life forms become the One.

In this African saga, One is the representation of the All. It can become two and the path is paved for the many that are to come. If these values are related to earth life forms, then the order of existence is realized and Flora becomes the earliest life form on this planet. It existed as grass, weeds, plants, flowers, trees, and shrubs of vegetation. It is the All of this in, One earth species.

Perhaps the best example is the concept that involves the Tree of Life. It begins as one root that grows on one stem. It grows as one upwards towards the sun. It will seek the sun untiringly, locked in this pattern of growth until it is cultivated. A simple snipping of the bud produces two branches that in turn can become four and then eight, to sixteen. With continual cultivation the possibilities of growth become infinite. This is the master plan for all earth life and is a free gift from RA, its star of billions of billions of years.

Chapter 24
The Power of Trees

The understanding of energy and of how this energy is found within nature allowed the ancient Egyptians to develop peculiar ideas that are not to be found in other ancient cultures. In most ancient African cultures this energy was most evident in plants and things that grow and develop. This was especially so with trees.

According to Budge, "In many isolated places within Southern Nubia and the Eastern Sudan are trees which men regard with reverence, but this may be the result of contact with the natives of Central Africa where people pray to Trees on certain occasions, believing that the spirits that are supposed to dwell in them can bestow gifts (or protection) upon those whom they regard in favor and ensure safety both to themselves and their animals when traveling."

The older a culture is, the stranger its customs. Environmental adaptation allows mental progressions that are visible in everyday habits, rituals, and routines that give definition to these ancient cultures. Thus a person or culture that is lacking this "antiquity" will most likely be unable to identify with customs that were practiced millions of years before them.

This is the case with the Hebrews, Greeks, Romans, and others that are the product of newly formed cultures, including the United States of America. In the annals of the order of the Great Ages, these are infant cultures. They lack the objectivity that is required when giving definition to beliefs that were derived in the Ice Age and before.

Africa was once known as the dark continent. This perception was the result of the inability of these outsiders to penetrate deep within its interior. All that was known to them concerning this interior was learned by contact with the outer African cultures. It was a land where no outsider had gone before. In this pristine interior that allowed life perhaps as far back as five to ten millions of years a go, the inhabitants developed a generic understanding of the value of plants and trees.

They found within them the magic of nature. Imagine if you will the arrival of mankind in this primeval setting of plants and trees, and things that creep and crawl, swim, fly, walk, and exist. From the plants they were able to find nourishment and within the trees they found shelter. This primary effect allowed these early hominids to identify with an order of growth and reproduction. It would seem that plants and trees became the reasons that culture first developed. They discovered back then that there is an energy of nature, evident within these forms of flora that can cause sickness, heal, or kill a person and this magic that they possessed was perceived as "godly."

The effects of trial and error are evident in establishing fundamental theories, dealing with an order of survival that is depending on vegetation. Does it taste good? What does it look like? Where can it be found? What color is it? Basic questions such as these allowed the discovery of nature and its two paths. They experienced that contact with certain plants and trees caused sickness, discovering that the cure for these inflictions were (usually) only yards away but on the other side of the path. Establishing an association of "good" and "evil" produced the disdain or the relevance that was shown.

A million years ago in central Africa, now called "Zaire," dwelled the ancient Bantu, a, "hemp"-based Pygmy culture, that some believe did coexist with the dinosaurs (crocodiles, alligators, etc.). Legend has it that they were masters of fire and the "ballast" and that they dominated the prehistoric creatures of this era. There was a presence of hemp visible everywhere and for good reason. This plant protected them with its power of intoxication, from the insects and beasts equally.

When grown in the correct manner it became a vegetational barrier of nearly impenetrable brush and shrub that provided them protection from the strongest and most feared of the dinosaurs. It also provided food, clothing, shelter, medicine, rope, and light. So it would only be natural that they developed an affinity with this plant, which grows from small seeds into huge trees annually

The power and the energy that is released by boiling the roots or eating the leaves or just mere contact with these plants and trees that could effect mood change or cause sickness and death was greatly sought after and highly valued. It is not strange to me that these Pygmies, who burned torches of hemp (and oil or animal fat) throughout the night, began to associate power to plants and trees. This was a culture that was able to separate the "trees from the forest."

Their order of gods of course begins with RA, and the others follow. What is unusual is the appearance of a new god that has only earthly domain, and his name was Hep. He was the god of the original burning bush. It was out of environmental necessity that Hep came to be and this process became the natural order of selection, for gods or goddesses.

Budge writes: "The meaning of the name of the Nile god, has not yet been satisfactorily explained and the derivation proposed for it by the priest in the late dynastic period in no way helps us; it is certain that Hep, later Hap, is a very ancient name for the Nile and the Nile god, and it is probably the name that was given to the river by the predynastic inhabitants of Egypt.

One of the oldest mentions of Hep that seems to relate him to trees is found in the text of Unas (line 187) where it is said, "Keep watch, o messengers of QA, keep watch, o ye who have lain down, wake up, o ye who are in Kenset, o ye aged ones, thou great terror, who comest forth from the Asert Tree."

It should now become apparent to the reader that the contempt that was expressed by the Hebrews, the Greeks and the Romans and even modern-day investigators (concerning the belief in Trees and Plants and the entire facet of Generic understanding that this culture related to, its Gods or Goddesses) was not warranted.

Instead it is a reflective reposition of the contempt (and the opinions that modern-day humans have for one another) and upon further review, it is found to be meaningless and totally unfounded. They have created problems that are feed from the rivers of doubt and the depths of their unknowns, regarding ancient cultures that are in order. They have in their own ways tried to separate the forest from the trees, never experiencing the jungle or the woods.

Chapter 25
The "Knot"

The concept of relating knowledge with the art of tying or untying knots begins with this hemp culture. Rope was introduced and its various forms of usage required the application of knots. There were knots that were used in the construction of shelter, weapons, tools, traps, lures, nets, boating, bridge building, etc. It can be said that necessity prompted the development of this art, of joining and securing and the linking together of ends that can serve one purpose or more.

Standing at three feet tall, the earth was closer to them than it is to us today. These Pygmies, although smaller in stature than we of today, were required to use more of the brain. Strange as this may seem, it is factual. Survival in primal times required an animal-like awareness that has not been present in humankind since the advent of fire.

The order of preparation is most natural, everything in its place and its time. A normal size person (five feet) would have been too tall and unable to correctly adapt to that environment. Mental superiority between we of today and these early Bantu people would depend on environmental conditions, with them being able to conform and adjust while our size would prevents us from conforming to Lilliputian proportionality. Fear of the dark and survival in the dark demanded more (alertness) expending of "Brain Power" and with the mastering of fire, this "fear factor" diminished and the parts of the Brain that were once on, began to turn off.

The advantages of size allowed the Bantu warriors to hunt beyond their haven of safety, that was provide by the Hemp growth (barrier) and allowed them a passage system that was only assessable to them. This Jungle of growth of vines, that creep and crawl, twist, and divide, seeking the Sun and creating natural knots that "highlighted," the mystery of survival.

The ancient Egyptians many times refer to the symbolic value of the knot. It was a concept that was known to both rich and poor. It stood for the idea of union or the bridging of two points that are separated: the joining of the one with the many and the all.

From the text of Unas: "Unas is the master of the offering and he tieth the Knot and he provideth meals for himself." In this context the Knot is the union between the world of the living and that of the dead and bestows upon the king the mystery of its power in both worlds. The dead pharaoh wants for nothing.

Perhaps the esteem that was shown to the Knot is best depicted in the fifth-dynasty painting of the two Nile gods, Hapi (god of the south) and Hapi (god of the north) in which the Nile gods are seen tying in a knot the stems of the lotus and papyrus plants around the emblem of union.

In this regard humanity may be considered the living knot of Seb and RA and the destiny of the one will become the fate of the All. Could this have been the message that the Persian high priest was unable to deliver unto Alexander and Aristotle?

"THE RAUSAURIAN ANK"

It announces the "Age of Aquarius" and hte revelation of the "COSMIC EGG"

Chapter 26
Hymns to RA

From *The Gods of the Egyptians* (p.5); The text reads:

Homage to thee, O Amen Ra, who dost rest upon Maat; as thou passeth over the heavens every face seeth thee. Thou dost wax great as thy majesty doth advance, and thy rays [shine] upon all faces. Thou art unknown, and no tongue hath power to declare thy similitude; only thou thyself [canst do this]. Thou art one, even as is he that bringeth the Tena basket. Men praise thee in thy name, and they swear by thee, for thou art lord over them. Thou hearest with thine ears and thou seest with thine eyes. Millions of years have gone over the world, and I cannot tell the number of those through which thou hast passed. Thy heart hath declared a day of happiness in the name of "Traveller." Thou dost pass over and dost travel through untold spaces [requiring] millions and hundreds of thousands of years [to pass over]; thou passest through them in peace, and thou steerest thy way across the watery abyss to the place which thou lovest; this thou doest in one little moment of time, and then thou dost sink down and dost make an end of the hour.

Homage to thee, O RA, thou lord of Maat, whose shrine is hidden, thou lord of the gods; thou art Khepera in thy boat and when thou didst speak the word the gods sprang into being. Thou at Temu, who didst create beings endowed with reason; thou makest the color of the skin of one race to be different from that of another, but, however many varieties of mankind, it is thou which maketh them all to live. Thou hearest the prayer of him that is oppressed, thou art kind of heart unto him that calleth upon thee; thou deliverest him that is afraid, from him that is violent of heart, and thou judgest between the strong and the weak. Thou art the lord of intelligence, and knowledge is that which proceedeth from thy mouth. The Nile cometh at thy will, and thou art the greatly beloved lord of the palm tree who makest mortals to live. Thou makest every work to proceed, thou workest in the sky; and thou makest to come into being, the beauties of sunlight; and the gods rejoice in thy beauties and their hearts live when they see thee. Hail RA who art adored in the Apts, thou mighty one who risest unceasingly. [p. 8]

Hail, thou Form who art One, thou creator of all things; hail, thou Only One, thou maker of things which exist. Men came forth from thy two eyes, and the gods sprang into being at the issue of thy mouth. Thou makest the green herbs whereby cattle live, and the staff of life for the use of man. Thou makest the fish to live in the rivers, and the feathered fowl in the sky; Thou givest the breath of life to that which is in the egg, thou makest birds of every kind to live, and likewise the reptiles that creep and fly; thou causeth the rats to live in their holes and the birds that are on every green tree. Hail, to thee, O thou who hast made all these things, thou Only One; thy might hath many forms. Thou watchest all men as they sleep, and thou seekest the good of thy brute creation. Hail, Amen RA, who dost establish all things, and who art Atmu and Harmachis. All people adore thee, saying, Praise be to thee because of thy resting among us; Homage to thee because thou hast created us. All creatures say, Hail to thee! and all lands praise thee; from the height of the sky, to the breadth of the earth, and to the depths of the sea thou art praised. The gods bow down before thy majesty to exalt the Will of their Creator; they rejoice when they meet their begetter, and say to thee, Come in peace, O father of the fathers of all the gods, who hast spread out the sky and hast founded the earth, maker of things which are, creator of things which exist, thou Prince. Life, health, and strength [to thee!], thou Governor of the gods. We adore thy Will (or, souls) for thou hast made us; thou hast made us and hast given us birth. [p. 10]

The awareness of the Cosmic Egg awakens the AK, BA, and KA.

Chapter 27
The Origin of "God"

The origin of "God" begins with man. Since the ancient Egyptians were the first Homo sapiens to assume, (this man) title they must be credited with the very first ideas about or concerning "God". "God" is not only known to humankind, but to all Earth forms of life and it is possible, that while observing these lower forms paying homage, to an unseen force or power, he (Hominids) was enlighten and adopted from them. It was known that Apes and Baboons gather together and in their own way and sing hymns, of praise to a rising Sun and to the fullness of the Moon. That which is most primitive to all Earth creatures became the basis, upon which the ancient Egyptians perceived and understood the realization of "God."

From the "Book of the Dead:" "God is One and only (RA) and none others exist with Him." This was the ancient Egyptian concept that was already millions of years old, before it, was adopted by the Hebrews. It continues: "God is the One, the One who has made all things. God is a spirit, a hidden spirit, the spirit of Spirits, the great; Spirit of the Egyptians, the divine Spirit. God is from the beginning, and he hath been from the beginning. (RA is sSelf-Created.) He hath existed from old, and was, when nothing else had being."

This is a mirrored reflection of the concept of "Intrafarance." This understanding was impossible for the Hebrews to comprehend, for it took millions of years for the ancient Egyptians to develop it and the Hebrews were only in Egypt for a little over three hundred years. As a result this depth is missing when the Hebrews equate "Oneness" with their deity and explains why they were unable to describe any essence of His being.

"He existed when nothing else existed, and what existed He created after He had come into being. He is the Father of beginnings. God is the eternal One, He is eternal and infinite, and endureth forever." This is more than an equivalent of being the "Alpha and Omega." This hymn relates to and identifies the need that humankind has, for "God."

"God is hidden and no man knows his form. No man has been able to seek out his likeness; He is hidden to gods and men and he is a mystery to all his creatures. No man knoweth how to know him. His name remaineth hidden; His name is a mystery to his children. His names are innumerable, they are manifold, and none knoweth their number."

The naming of (a) God that has a hidden name, that is unknown, is still practiced today. Does it make sense? If it does, this is where it begins, in scriptures that redundantly profess the attributes of God and seemingly in this way, establish a personal relationship with him. "God is truth, He liveth by truth. He feedeth thereon. He is the king of truth and he hath established the earth thereon. God is life and through him only man liveth."

It seems that even in those primitive times, it was known to the "Hominids" and later to "Homo sapiens" that man cannot live by bread alone. Somewhere there must be God if our lives are to have meaning. This was the understanding of God that was and is lacking in other cultures and it was replaced with sacraments and rituals that are often meaningless. Its expresses that God is truth and that the earth was meant to be a place of truth and as we now know, that truth has much to do with (the truth) that concerns Stars.

"He giveth life to man, He breatheth the breath of life into his nostrils. God is father and mother, the father of fathers and the mother of mothers." Not many religions are able to deal with the fact that God is both male and female. If the source of all life is not everything that "is," than it is not the source of all life. In this book we relate to the "All of Alls." this reflects the reasoning that "One" being can be "Two" male and female. Or that one can be three (light=solid, liquid, and gas) or that one can be seven (White light). The true Spirit of God is only found by this understanding that "One" is only a part of the "All" and vice versa.

The hymn continues: "He begat but was never begotten; He produceth but was never produced; He begat himself and produced himself." It would not be far fetched to think that if this culture was dominated by women, that the text would read: "She begat herself etc." In this way the reader can see the reason behind the Gods and Goddesses. In very ancient Egypt the women were equal with men, in many regards and would share wealth and power equally. But it was thought that if any of the other existing animal forms could alter itself and self create itself anew, the form, that it would choose is that of "man" who is the spirit of copulation. This earth is a story of the truth and in this aspect men and women are not equal.

"He createth, but was never created. He is the ruler of his own form and the fashioner of his own body. God himself is existence, He endureth without increase or diminution, He multiplieth himself millions of times. He is manifold in forms and in members. God hath made the Universe and all that therein is. He is the creator of what is in the world and of what was and of what is and of what shall be. He is the creator of the heavens and the earth and of the deep and of the water and of the mountains. God hath stretched out the heavens and founded the earth…. "What his heart or mind conceives, straightway came to pass." When he hath spoken, it cometh to pass and endureth forever. In the Hebrew scriptures, God is known to create the heavens and then the earth, but it takes seven earth days.

"God is the father of the gods. He fashioned man and formeth the gods. God is merciful unto those who reverence him and he heareth him that calleth upon him. God knoweth him that acknowledgeth him, He rewardeth him that serveth him and he protects him that followeth him." The ancient Egyptians, believed, that, everything that came from God carried a portion of God, within them (It) and even the lower forms of life enjoyed this distinction. Thus that part of God, that is within you is Godly. And it is seemingly visible to all of the lower animals (and life forms) in nature except humankind.

Upon careful review, the reader will find science and religion and the essence of us all, in these hymns to that very special star, called RA that has existed for billions and billions of years. All earth species are covered in these doctrines of praise and without discrimination towards any earth creature. This is the understanding that scholars of the past, that did not use the stars as a grid for these doctrines were unable to arrive at. This is what is meant by Monotheism with substance!

The world as it was in the time of Nebuchadnezza.

Nebuchadnezza and his advisors, relishing in the victory over the Egyptians, entered the "Holies of Holies" (of) the inner temple. This place was only for, those, special priests that possessed, a unique knowledge of "RA" and the "Stars." The "great king" was "dirty" he and his entourage entered, wearing their moustache's and long beards. The priest that performed, duties, "there," were all, clean shaven, every where their bodies were devoid of hair. In his ignorance he had violated the sanctuary of a "God" that has "existed" for "billions and billions" of years. For this he was cursed, he was told by the "High Priest," that even a "great king" can began the year as "ruler" and end the year as a "beggar" and that because of his inner temple violation, this would be his destiny.

Nebuchadnezza was enraged and ordered the destruction of the temple and the "priest hood" and he killed many that were the "messengers" of this "God" of "billions of years." Of those that were in the employ to this special place (Holies of Holies) he took into captivity 300 people and they were carried off, to a city called "Babylon."

It was customary for the affluent ancient Egyptians to send their children to study in the outer temple, for even those who were of the Great Line of Pharaohs were known to be instructed there. In the whispers that pass in the night, they learned of their fall from prominence into captivity. What happened to these three hundred people? Did the knowledge of "billions and billions" of years just fade away? In its own way this book is the accounting of that knowledge and of those three hundred people who shared a destiny that was linked to two "great Babylonian kings," the modern-day world and the God of "billions and billions" of years.

Nebuchadnezzar had unwittingly carried the seeds that would be scattered throughout the length of the earth, being dormant in this age of darkness (Pisces) to which he and his father were the harbingers. In a dream he was visited by a "Demon of the Light" and was shown the future, from his time until the coming of the new age (Aquarius). He viewed the carnage and ignorance that would befall the Earth because of their religious and territorial ambitions. His mind was overwhelmed with sights that revealed to him the technology of the ages to come and the more he understood, the stranger he acted, until he was driven mad through the power of "enlightenment." Before that year was ended, He, the great king, had gone mad and become a beggar.

This knowledge that involved the existence of a god of billions and billions of years was hakka, known only to a handful of priests and pharaohs. These priests and kings believed that they were of this god's Issue (member or pallus) and that they were of the divine line of kings, his appointed servants, and keepers of his trust and sworn to this sacred duty at birth. Since the appearance of mankind upon the Earth, there have always been those who were chosen to do its bidding, beginning with Ausaures. The earthly presence of this god was known to dwell within the Holiest of Holies of the inner temple, the most "pure" place on the planet. Because of this intervention, the first would now become the last!

These three hundred captives represented what was left of the priesthood and royalty of Egypt and the diminishing of the power of their ancient gods. For this reason, they were valued by Nebuchadnezzar. It was a custom in those days to take into captivity the sons and daughters of the affluent and royalty. His father Nabopolassar enjoyed taking into captivity those that served other gods, especially those that were not of his liking. For him there was only Marduk and the Seven Tablets of Creation, and his desire was that any other theological theorem should be destroyed. By the capture of its priests, holy men and messengers he believed their religion would die out (vanish). And that in this manner, even the gods can die.

Around the year 600 B.C. (or 666 B.C.) the dawning of the Age of Pisces, the Assyrians, Babylonians, Egyptians, and Hebrews are the main players that were (are) the cause and effect of today's religious beliefs. In the text of this book I go into depth concerning the ancient Egyptians and their gods, and also I mentioned that this was a time when the gods were at war with one another. Now let us briefly consider the beliefs of the Assyrians and the Babylonians and the Hebrews of this period.

The "Seven Tablets of Creation," Which were written in cuneiform and contained the views and beliefs of the Assyrians as to the origin of the gods, the world, and mankind, were also "scripture" for the Babylonians and Hebrews. These Seven Tablets of Creation introduce their gods and goddess and pronounce an understanding that is borrowed, but not completely understood. From this foundation of religious beliefs that are not indigenous to ancient Mesopotamia, the Assyrians and Babylonians found the audacity to defile, all the other gods. If the reader is vigilant they will see the duplicity of religious theology that perpetuates a "mandala" of forms in which creation and God are introduced, but not fully explained.

E. Wallis Budge writes in *The Gods of the Egyptians*, Vol. 1: "The old company of primeval gods mentioned in these tablets are also eight in number, and they fall readily into four pairs. The first consisted of Apzu-rishtu, i.e, "the primeval abyss," and Mummu-Tiamat, the mother of offspring by him." He states that "These two deities then represent the male and female powers of the watery mass, which contained the germ of all life, and of every kind of life, and they existed at a time when, of the gods, none had been called into being and none bore a name and "no destinies were [ordained]." When "their waters were mingled together," then the work of creation began. (If we relate the "waters" to a black hole, then the creation of the light is revealed in theme.) Budge concludes that Apzu-rishtu and Mummu-Tiamat are the exact equivalents in the Babylonian cosmogony of Nu and Nut in the Egyptian and that they are the originals of the Greek gods Zeus and Hera. (Is this duplicity?)

They lacked the knowledge of that special ingredient (radiation) that is the true source of all creation, and their understanding of the light was so limited, it is possible that they were unaware of the true meaning of their own religious doctrines. At this time the Hebrews and their one god (monotheism) had no knowledge of the origin of light or radiation. This is evident in their scriptures then and now. In all fairness it would be impossible for a culture that was born under the sign of the Fish to comprehend the true understanding concerning the origin of the universe.

Budge says: "The Assyrian copy which we now have was made during the reign of Ashur-bani-pal, king of Assyria from 668 B.C. to 626 B.C. presumably from a Babylonian archetype of an unknown period…. It is surprising therefore to find so much similarity existing between the primeval gods of 'Sumer' and those of Egypt, especially as the resemblance cannot be the result of borrowing. It is out of the question to assume that Ashur-bani-pal's editors borrowed the system from Egypt, or that the literary men of the time of Seti I borrowed their ideas from the literati of Babylon or Assyria and we are therefore to the conclusion that both the Sumerians and the early Egyptians derived their primeval gods from some common but exceedingly ancient source.

"The similarity between the two companies of gods seems to be too close to be accidental, especially as there is every possibility that the Sumerian system was taken into Egypt by the same people who carried into the country the art of making bricks, the use of the cylinder seal, and the like." Be this as it may, it is certain that the company of primeval gods, which we have seen, that was common to the Sumerians and Egyptians, was quite different from the companies of gods of which Ausaures and Ra-tem were the heads, in Egypt, and also from those which were formed in Babylonia and Assyria when those countries were inhabited by Semitic populations.

The Hebrews who originally lived in Asia-southern Babylonia-were also greatly influenced by these Seven Tablets of Creation. Budge writes: "It is a curious fact that the Hebrews, who borrowed so largely in their cosmogony from Babylonian sources, did not also borrow in some form or other the monster Tiamat, which played in their mythology the same part that Apep or Apepi played among the Egyptian gods."

Everywhere we look while reviewing the past, concerning religion, we find the views of the ancient Egyptians. The Seven Tablets of Creation are unfulfilling, as they are duplicates of Sumerian and Egyptian mythology. These great civilizations mirrored this knowledge that today is professed as the "gospel." There is no scientific inkling of the real forces of the universe as we understand them today, light and radiation. The knowledge of the stars that have a life span of "billions and billions" of years is absent. As these civilizations were ignorant of this, they were unable to borrow or seize it.

The Persians, Greeks, and Romans are the recipients of this understanding that comes to us out of Asia and they passed on these faulty duplications of religious misconceptions. They all found it advantageous to invade Egypt and to borrow from them religious concepts that were beyond their scope of reasoning. The order of existence was unknown to them all. Science as we understand it today is not reflected in their theories of origin.

For nearly 2,600 years the earth has been without a "Holiest of Holies" and humanity has suffered greatly because of this. We as a species have been led by the blind leading the blind. As a descendant of the three hundred people who were held in captivity in ancient Babylon and in the making of modern-day America, once again comes the story of the Star-God called "RA," who is the lord of "billions and billions" of years and the source of life for a planet known, as Seb.

Chapter 28
Nebuchadnezza's "Dream"

Rodolfo Benavides writes in *The Dramatic Prophecies of the Great Pyramid*, "The Old Testament episode of Nebuchadnezza's dream is a helpful illustration of the threat, the Jews of the time of Christ had bought down upon themselves, though they were neither the first nor the last people in history to do so. 'Nebuchadnezza, the rich and powerful king of Babylon, dreamed of a gigantic statue whose head was of gold, its breast and arms of silver, its belly and thighs of bronze, and its lower legs and feet of iron and clay.' A stone coming from "nowhere" strikes the statue's clay feet. The statue falls and is broken up into thousands of pieces, which in time, turn to dust and are blown away by the wind. The stone, which caused the change, meanwhile grows into a mountain."

It was shortly after his return from the conquest of Egypt, that Nebuchadnezza began to have this dream. For a series of nights the dream would return to him and each time that he awoke from this bewildering dream, the details of what he "dreamed" were gone from him. Ancient Egyptian literature is abundant in stories that involve the dreams of the great pharaohs, considered the primary way in which the gods speak to priests, kings, and mankind and make known unto them their desires. It was the desire of the God of "billions and billions" of years that Nebuchadnezzor be driven mad and that the fulfillment of the curse come to pass. This dream represented the beginning of his madness.

In the dungeons where the three hundred were kept were other prisoners and holy men, gathered from around the then-known world. Among this large gathering of "captives" there were Hebrews or this cult of Semitic people that claimed to be from the line (pallus) of just one man called "Abra-Ba-Hama." These people were the "Jews." It seems that they had fallen out of favor with their "One" unseen God and their nation of Israel also suffered the ravishing of the great king. These cultures represent the players that willingly did the bidding of their deities. The gods were at war with one another and mankind itself.

There was a great disdain for humankind that had looked away from the gods, truths, of their actual meaning and their existence. Polytheism, or the Asian concept, as practiced by the Assyrians and Babylonians, angered the old gods. These new kings had allowed the ancient temples and monuments to be left in a state of disrepair and were not supportive of the needs of their own peoples. The cries of the young ones and the old people, their groans of hunger, moans of pain, and despair, heralded the misery and the hopelessness that abounded within these kingdoms and as these sounds ascended, upwards, into the heavens their shrieks of anguish angered the gods. Everywhere in every country the rulers of these nations were found to be wanting. This was also the case of Egypt and its ancient "gods" and why it was allowed to be vanquished by the Babylonians, and the esoteric purpose of the captivity of the three hundred.

The fate of the gods, the world to come, and humanity would now be decided, in this ancient' "pearl of the desert" city called Babylon. Its outcome was revealed to Nebuchadnezzar in the nightmare he repetitiously dreamed. He called together his astrologers and magicians and bade them to pray-tell the interpretation of his dream, but they could not. For how could they interrupt a dream if they did not know the substance of the same? With his sanity waning he ordered the execution of all the captives within the city, not realizing that this would be his last official order as the great king. In the eleventh hour a Jew was found; his name was Daniel and he made known the meaning of the dream unto the great king and his court.

"You are the head of gold," he told the king. "After you shall rise another kingdom, inferior to yours and then a third kingdom of bronze, which shall rule over all the earth. And there shall be a forth kingdom also, as strong as iron (but) because iron breaks into pieces and shatters all things; and like iron, which crushes, this fourth kingdom shall break and crush (all) others. And as you saw the feet and toes of the statue were partly of potter's clay and partly of iron, it shall be a divided kingdom. Some of the firmness of iron shall be in it mixed with the miry clay, but they will not hold together, just as iron does not mix with clay.... And in the days of those kings the God of heaven will set, up a kingdom that shall never be destroyed, nor shall its sovereignty be left to another people. It shall break up all these kingdoms and bring them to an end and it shall stand forever."

This dream foretells of a kingdom that will be established by the "Great God" and his chosen followers, unlike the kingdoms that are established by mankind. It hints that the nations and the beliefs of these nations were not in accordance with the laws of this "Universal God," that is the true "One Creator" of all the gods and all matter. Suggesting that the, Assyrians, Babylonians, Egyptians and Hebrew cultures of that era were all in disorder.

At this time Egypt was in the control of the (Semitic) "conqueror" Seti 1(666 B.C.) and his offsprings. However they were outsiders and not truly understanding of the "ancient religion" of Egypt. Regarding the Hebrews, Benvides says: "Perhaps the Jewish people of this time had strayed from the path of serving the Lord God Jehovah, to dedicate themselves to the worship of the Golden Calf. They therefore; had given up their right to be the people, singled out as mankind's spiritual guide, as with the Assyrians and the Babylonians. In this regards the year 666 B.C. is surely the time of Satan and his followers, and when first they came into power.

From this well of unrighteousness, the nations that followed have drunk. They passed on the doctrines of this the darkest era in the annals of humankind, and the (our) world is the worse because of it. Our hope comes in a nightmare of a dream that speaks of a time and an age (Aquarius) when mankind will once again come to understand the real meaning of a God that has existed for "billions and billions" of years.

Persian, Greek, Roman, and British Empires have come and gone; only one remains. I am a descendant of the original three hundred, whose families and people (the seeds that were scattered around the world) have suffered captivity in that fourth kingdom which is modern-day America, the land where clay and iron are mixed but do not cleave to one another. Nebuchadnezzar's dream depicts the racism and separatism that is the "gospel" practiced by white Christian America religiously, to its exact likeness. I have compiled a collection of thoughts that attempt to explain our meager existence during our period of captivity in the greatest of all the kingdoms of mankind, North America or the good old U.S.A.

"THE HIDDEN ONE' Emerges from the "Cube of Existence"

Chapter 29
The Great Line of Kings

This is the story, an unholy story of the captivation of a "holy" nation. The story of the subjugation and the ruination of this holy nation, that was born from the great line of Kings.

And how without their past, the first became last and how the story of their glory, to be born from the great line of kings became a story so gory that in their minds, captivity made them a nation of slaves. Captivity reduced them to slaves. In captivity, the nation became slaves.

From diamonds and gold, emeralds, and numerous things, natural possessions of the great line of kings. Everyday trinkets to the great line of kings but captivity made them a nation of slaves.

Because without their past, they lived life in halves-half man, half beast, half slave, and this trend of thought was all that they sought, all that they bought, without knowledge of yesterday. Without the knowledge of yesterday.

Do you know what it feels like to be an uncrowned king? Lacking the honor and glory, health and wealth and just about everything? Well, listen to this story, this unholy story, about the captivation of a holy nation and the ruination of this holy nation that was born from the great line of kings. We were born to a great line of kings.

The "Man" developed a plan, a "Master" plan, a "Sethian" plan; in it, he needed the African black man. He needed the black man. Great slave ships set sail, acquiring him became a great deal, cause he needed the African black man. A great demand there was, cause he needed the African black man.

They purposely eliminated our native tongues, they did this while the country was still young, when the original natives still owned their land, we were captured and taken away from our desert sands. Taken against our will from our desert sands and forced into captivity, in a faraway land. We were forced into captivity in a faraway land.

Yes, in those days we fit into the plan of the Man. An unholy plan for this unholy land. We would help steal, wheel and deal, from the Natives this land, if that is what Master commands. We were compelled to do all that Master commands. This was the reason we came to this land. Must do all that Master commands. All that Master commands.

In those days we fit into the plan of a slave-owning man. We fit into the plan of the Man. Into an unholy plan, in an unholy land, to do hard labor at Masters command, the reason we were brought to this land. We worked our Palluses off in this unholy land, did this at Master's command, we who were born to the great line of kings. We were born from the great line of Kings.

We fit into the plan of the Man, the slave-owning Man. We fit into a plan that was his Master plan and back then we got off on it, escaped into It, accepted (religion) from it, our past was lost within it. Fit well, we did, into the plan that allowed the captivation of this holy nation, that induced the ruination of this holy nation that was born from the great line of kings. We were born from the great line of kings.

We became part of a plan, that unholy plan that allowed the captivation, the subjugation and ruination of this holy nation that was born from the great line of kings. From power, diamonds, and gold, into slavery we were sold, where we did what we were told, those born to the great line of kings. Those born from the great line of kings.

Yes, in them days we were needed by the man, with this unholy plan, we were needed by the man and we were in great demand. We did labors that no others could do, labors that some refused to do, labors a man should not be asked to do.

The only ones performing these labors has a name and that name is Who. And they became the black niggers, the producers of me and you. These black niggers fit into the plan of the slave-owning man. That unholy white Man with the Master plan, for the ruination of this holy nation that was born from the great line of kings. For the subjugation of this holy nation that was born from the great line of kings.

We fit into his plan, we are still fitting into his plan, and black ass niggers are still getting off by fitting into the plan of the slave-owning white Man, that allowed the captivation of our holy nation. That was born to the great line of kings.

Way back then, way, way, way, back then, we were needed by the Man, way, way, way, back then. They needed old black women, young black women, young boys, young girls, and young men. They even needed the older black men, way, way, way, back then. Yes, we were all needed way back then, needed by the Man, way back then.

But when this need had ended, we blacks were not commended and found ourselves non-friended. They developed a new plan, a genocidal plan, developed this when the labors had been done, back when the wild, wild, west was not yet won. They wanted us to go away, did not want us to stay in this unholy land that was gained in an unholy plan by a killing Man with an unholy Master plan.

Yes, they developed a genocidal plan, to free this unholy gained land of them-old black women, young black women, young boys, young girls, and young men. A masterful plan that will free this land, that will rid this land of them. Especially the older men, they must get rid of them, for the older men claim lineage from the great line of kings. They claim lineage from the great line of kings.

The older men still remembered when they were not a part of this plan, this unholy plan that allowed the captivation, subjugation, and ruination of this holy nation that was born from the great line of kings. The older men still remembered when we were part of the great line of kings.

Way back then, way, way back then, kidnapping black people was legal way back then. The raping and taking of black women was legal way back then, the bondage of a nation was legal way back then. Did what they want to do to niggers and it was legal way back then. The captivation of this holy nation was legal way back then. All this shit was legal way back then.

The law, it was blinded. The truth, they could not find it. They all seemed one-minded, towards the captivation and the subjugation and the ruination of this most holy nation that was born from the great line of kings. They enslaved those that were born of the great line of kings. They were about the separation and decimation of this holy nation that was born from the great line of kings.

Hanging black men was a fancy way back then, while we were in captivity. Hanging black men was a fancy way back then, the crippling of black men was a fancy, way back then, the Man was not nice way back then. The plan was not nice way back then, this unholy plan of the Master Man, was all telling way back then. All too revealing way, way, way, back then, it meant the taking of this holy nation, the separation of this holy nation that was born from the great line of kings.

We blacks were acquired way back then, seized, caught, mugged, and bought way, way, back then. This was the will of the Man way back then, it was the law of the land way back then. Our holy nation was stunted way back then, our culture was plundered way back then, and our lives and religious beliefs were interrupted way, way, way, back then.

And when the Natives lost their land, we were part of that unholy plan. We played our part in that unholy plan, that was developed by an evil Man, that truly did not understand that he was part of evil's plan, the Master that evil commands, and we both suffered greatly at his hands.

We are the iron of captivity. The slave-owning Man is the clay. America is the kingdom. The era is "today. Captivity was the way that this land first employed the hands of those born to the great line of kings. The unholy plan paid for our tickets, they fronted our bill to bring us here against our will, and then treated us mean and wicked. We who were born from the great line of kings, born from the great line of kings.

With Ausaures began the great line of kings. Heru his son was from the great line of kings. Aspelta and King "Tut" are from the great line of kings. Rasaures is born to the great line of kings. All blacks, we are brothers, so stop killing one another. We are all born to the great line of kings. We are born to the great line of kings.

We are yesterday's captives and tomorrow's Masters, and once again, on this Earth, the last will rise to first, for we are born from a great line of kings. From the first to the last, that was our past, or, way, way back then. When the last becomes first, the new age will burst and that's when our futures, begin. For we are born to the great line of kings, born to the great line of Kings, we are born from the great line of kings.

"THE BENNU BIRD"

It was believed this "Avis" flew from the Sun (RA) to the Earth (SEB) upon landing
it incineration itself and from its ash earth life was produced.

Chapter 30
The Pharaonic Order

It was not by luck that the ancient Egyptians developed their understanding of the universe. To begin this understanding, we are required to travel backwards in time, possible five to ten millions of years ago. At that time the first hominids (mankind's earliest ancestors) were known to roam the continent of Africa. They had only one thought and that was survival. As modern-day anthropologists continue to unearth their remains, the truths of yesterday become the lies of today.

The first black man on this planet was a hominid and his name was not Adam. He was called "Ausaures," and all the black peoples in the world today are of his lineage. Ausaures says: "I am yesterday, today, and tomorrow, the lord of millions and millions of years." The story of Ausaures and his descendants was purposely deleted from the books; along with the accomplishments of his children and the many great civilizations that they introduced for the benefit of all of mankind. Why was he so special? Because he was instructed by the self-created god Thuti and was wise in the usage of words and had the ability to give the word written form. It was believed that after the earth was stocked with life, Ra commanded Ausaures to go about the earth and give names to every creature that existed there and to make a written report. After he had done this he found great favor in the eyes of RA.

Before Abraham and his offspring, before the Chaldeans, Babylonians and Sumerians, there was Africa and a nation called Egypt. Ausaures and his children claimed this place as home, and so millions of years ago Africa's inhabitants were developing the doctrines of the light. Ancient Egypt allowed the rapid advance of knowledge that developed writing, art, science, and religion under one uniformed order. It was the first system to understand the need for schooling. These instructions began in their childhood and continued for some their entire life. It was necessary, for survival demanded knowledge.

The understanding of natural occurrences and the need to construct shelters, evident back in those early times, developed into an African pageantry that is revealed in their tombs, temples, and structures. From these Pharaonic images comes the grandeur that is witness to this amazing culture. The pyramids, giant structures that tower high above the earth, and relate to an age that demanded constructive development, are, also the results of these doctrines of the light.

This Pharaonic order introduced into the world the wonder of majestic masonry that is a direct result of the doctrines of the light. Masonry in ancient Egypt was based on the bounding of the body, mind, and soul, producing a being that was adept at solving the mysteries of the light. To the ancient Egyptians, light represented the creative principle of the universe. As their understanding of this form of matter increased, science was introduced and religion was born. The religion of the light perpetuated the need for man to continuously better himself.

By informing his son or daughter, he prepared each to master more of life than their fathers or mothers. In this manner, a son becomes greater than the father and a daughter wiser than her mother. In this way they created an ever-changing and developing culture. This was the introduction of "progress." They believed that Ausaures was their paternal father and that Ast was their maternal mother. Their son Heru produced four sons, Hapy, Taumutet, Amast, and Qeubsunub. These were the original people that populated ancient Africa and ancient Egypt. These beings became the nucleus that produced the brotherhood and the sisterhood of the light. Two million years later, because of the discoveries of anthropology we are able to observe this African culture that was dedicated to producing men and women of higher intellect.

Chapter 31
How Sweet Is the Truth

The master craftsman and teacher superior was Thuti. He was the personal instructor of Ausaures, developer of the sacred writings and the medium between the gods of light and humankind. The Great Pyramid of Giza, is symbolic of the sister-and brotherhood, for it incorporates brick masonry and free masonry into one structure of divine origin. Along with its neighbor the Sphinx, they remain the only wonders of this era still standing. It is therefore only proper to pay homage to every aspect of their construction.

These two structures are symbolic of the ideas and the theories that relate to the stars. In this world and for the world that is to come were the practioners instructed, for man is a creature of these two worlds. The two eyes of Heru are symbolic of the world of the living and the world of the dead. Numerically they represent decimals, fractions, percentages, and esoteric symbolism.

Ancient Egypt was a country that was primarily farm land. These farms naturally tested and developed the Egyptians' minds at the earliest of ages. When one had reached the age of acceptance, his next instruction would be provided by the outer temple. Fresh from the farm, with young minds that are responsive to laws that apply to husbandry, they were now ready to receive instruction in the value of numbers, writing, language, and music, astronomy, architecture, and morality, all of which were incorporated into their religion of the light. This was a gradual process that often spanned a lifetime.

Chapter 32
Candidate School

The apprentice was introduced to pottery, weaving, stick joining, meditation, astrology, proper diet, and the art of fasting. The diet and fasting allowed the mind expansion necessary to comprehend the difficulty of the studies. Only through proper diet and fasting could the apprentice say: "I am pure, I am pure." The meditation allowed the joining of body, spirit, and mind. Having reached this level of instruction, those who had advanced acknowledged that they "come in peace." Each time the candidates advanced a degree, they were awarded titles that denoted their advancement. There were thirty-four degrees attainable in this world and one that would be received in the next world or afterlife.

Each soul was destined to travel to the underworld, where accepted souls were reborn spiritually and continued onward in the plan called "the evolutions of the light." As one was able to comprehend, he was awarded and advanced. Upon reaching the seventh degree, he was designated the title of the "perfect master or scribe." His duties were then divided between solitary temple studies and performing the civic duties of a scribe, throughout the length of the land.

These duties were numerous: first and foremost, representing the law of the land. These scribes and judges were honorable men and for good reason; because of their instruction for one, and because the penalty for misusing their power was death. Any judge found to be corrupt was killed. Thus their rulings, e.g., in land disputes, etc. were generally arrived at fairly. Many candidates attained the seventh degree and found it rewarding enough to remain there their entire lifetime.

"THE YOUNG OSIRIS"

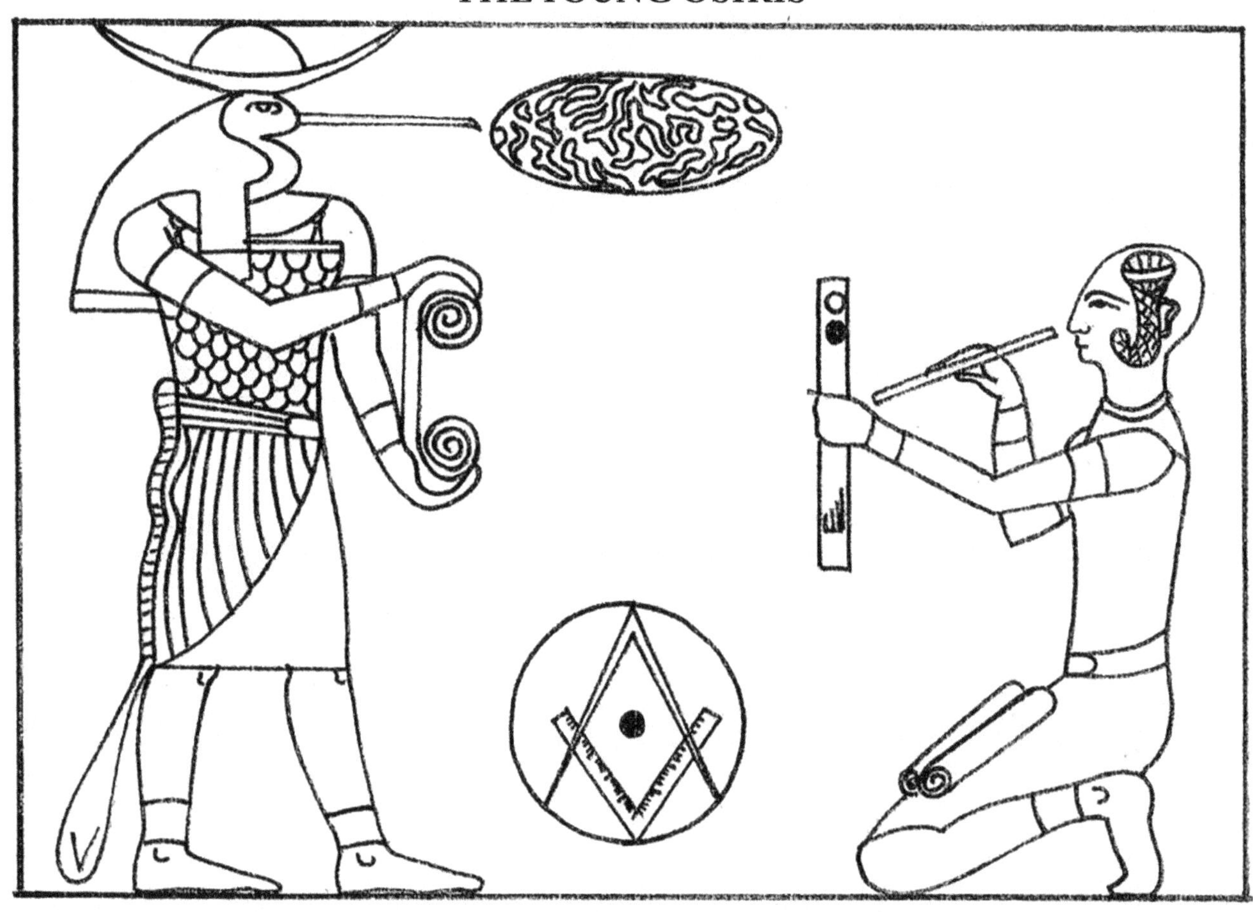

Instructed by "Thuiti" Lord of the "MEDUNETERS"

Chapter 33
Steps to Higher Learning

The prominence of the thirty-fifth course of masonry visible on the great pyramid itself esoterically reflects the power of seven. There are seven earth species, seven colors, seven sounds, and seven openings in the cranium. The ancient Egyptian word "seftex" is not so different from the English word "seven." This degree is known as the "Place of Maat." As the country grew and expanded, these temples produced the men and women whose minds allowed ancient Egypt to become the world's first great (builders of stone) culture.

Ancient Egypt was a nation of forty-two Nomes, each of which was governed by a lord or overseer who shared this position with a prince, princess, or prominent local figure. This cadre of royalty, priest, scribes, and local leaders developed the system that allowed the growth and organization of very ancient Egypt while other cultures of that period lagged behind. The fourteenth step was esoterically entitled: "Giver of Winds." Every seventh step, was esoterically named. Twenty-one steps represented the "Weight of the Place." The twenty-eighth step was known as the "Ox of Seb" and thirty-five was called the "Arm of Shu." Jesus, called the Christ, was the last known graduate of this order.

Chapter 34
Attainment of Higher Learning

This unusual order of black men and women that produced craftsmen, scribes and judges, architects, governors, and kings was the intended purpose of the Pharaonic Order. They established a standard that many other cultures borrowed from and one that is still in practice today. As the candidates advanced in degrees, in stages of seven, they were given the priestly duties and titles.

It began with the lay priest, who was considered the beginner, unfamiliar at the art of serving the gods and goddesses. The meditating priest seems to be more comfortable in this service, followed by others, each more adept and enlightened than those that preceded him. The "kneeling priest" was very effective in serving the deities, followed by the "attendant priest," whose duties involved supervision of those subordinate to him.

Ancient Egypt was a land where the people very much enjoyed festivals and ceremonies. At such times, the symbols of the gods and goddesses would leave the temples and go to the people, often parading through the city and countryside. This is the segment of the priesthood that performed that function.

The instructor priest was responsible for the curriculum by which the others were informed of their requirements. The priest was a symbol of the complete understanding of this order, outranked by only the high priest, who was the oracle between man and the gods and goddesses. The five water lilies on the chart represent the five races of mankind. The seven lotus flowers symbolize the seven continents that have produced all earth languages.

The two sisters Ast (Isis) and NebT Het [Nepthys] reflect the importance that the male Egyptians ascribed to their mothers, sisters, and daughters. Intelligence in their women was a desired attribute. Ast was accredited with the knowledge that enabled her to revive her husband Ausaures from the realm of the dead. They appear emerging from the Ankh, the indigenous symbol of life.

The children of Heru designate the Cardinal points. Taumutet, the jackal-headed one, is the East, the ape-headed Hapi, the North, the falcon-headed Quebsenuf, the West, and Amast, who has the head of a man, the south. Anpu and Amut are seen tending the scales, while lady Shait rests opposite the heart, assuring a true accounting.

The two boats of the sun, morning (Aton RA) and evening (Atum RA), revolve around our sun, centered, with the tools of masonry appearing on its face. Below, the high priest takes from the solar fire the flame of eternal life. Nut the sky goddess is shown within the head of the Ankh, while Ausaures the lord of millions of years appears in the hieroglyphic form. Other visual symbols are the new age Ankhs held by Ast and Nebt-Het and the familiar symbol of American freemasonry. Finally the discerning eye can discover the face of Heru esoterically staring back at the beholder.

"THE PHARONIC CHART"

THE PHARONIC ORDER

"ORION"

"THE DECON OF STARS"

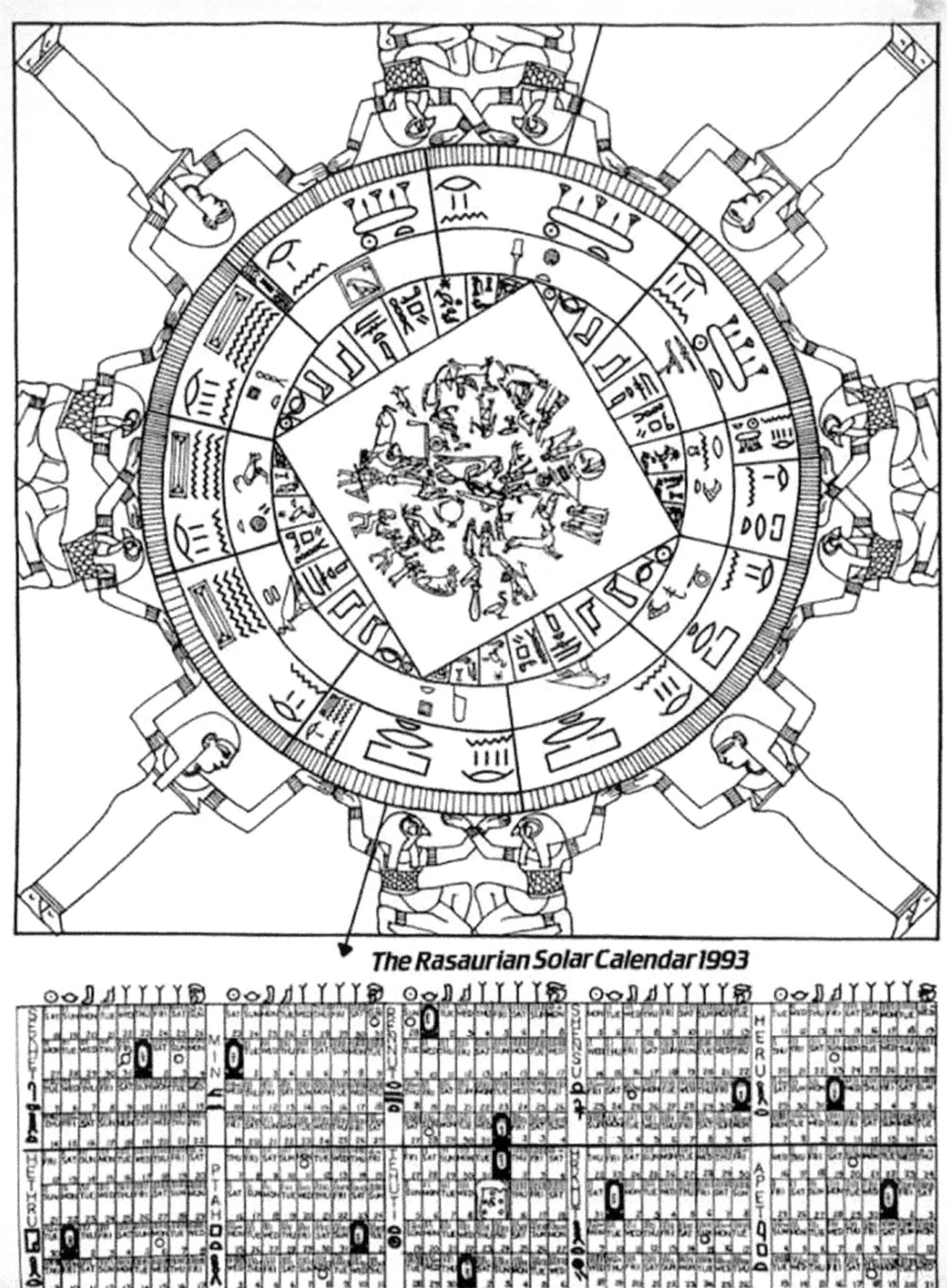

The Rasaurian Solar Calendar 1993

Chapter 35
The Rasaurian Solar Calendar

Before the Romans and the Greeks there were nine days in every week and
time was kept by charting stars that were visible on their calendars.

The knowledge of the stars is evident in the ancient Egyptian solar calendar, and the reader will again come to the realization that this understanding was omitted in the development of calendars of other cultures. In the text I have mentioned a few and will now elaborate on those calendars and reintroduce the ancient Egyptian solar calendar and their unique concept of star time.

According to *The New Book of Knowledge*, "A calendar is a system for keeping track of time. It is a kind of clock for the entire year. While a clock measures seconds, minutes and hours, a calendar measures days, weeks and months." This is the accepted definition of a calendar. In ancient Egypt the calendar was a seasonal time chart regulated and observed by understanding the procession of the stars.

From a section entitled; "How The Modern Western Calendar Measures Time" "The basic units of the calendar are the day, week, month and year. Our modern calendar has 365 days, divided into 12 months as well as 52 weeks (with one extra day): each week has seven days. These divisions of time are based on the movements of the earth, the moon and the sun. Where is the mention of Stars?

The ancient Egyptians Star dated everything. In the temple of Het-Heru/Hathor that was discovered by Napoleon's General Desaix during the French intervention of African Egypt in 1798 A.D. Inside on the ceiling of an upper room was found the "Zodiac of Dendera." Starting with the sign of the Lion (Leo) at the Spring Equinox, this circular zodiac indicates a celestial alignment that occurred over 9,000 years ago; (making) it the oldest and most complete of any astronomical charts ever found on the earth.

This Zodiac of twelve Astrological signs, written in an African dialect, demonstrate that it is the gift of superior black minds and supersedes accepted concepts that suggest the Chaldeans or Babylonians were the originators of astrology and astronomy. Or that the zodiac with its Latin names was the result of Roman efforts. (Zodiac in Latin means Zoo.) That famous "gap" always presents itself. Nine thousand years ago there was no Europe or Rome and there were no Hebrews or Greeks or Babylonians, etc.

The Zodiac is observed inside the square center of the Rasaurian solar calendar. The alignment also indicates a time when the star Sothis (Sirrus) rose cosmically with RA our sun, under the sign of the "Great Year of Pisces." This helical rising of two kindred stars occurs every 1,461 years and was in unison with the Epagamal days or the ancient Egyptian New Year; and was known as the "Symbolic Year" or the "Year of Years." Here was the ideal model of a 9,000-year old calendar that was in alignment with the stars and it also was rejected. Why?

Let us examine the information that is and was available. The Egyptians conceived of a cycle that included day and night and "Tuat" and "Duat," and these periods were awarded a certain amount of time. (In the modern calendar this cycle is called a day, the amount of time it takes for the sun to rise set and then rise, again. Each day includes a period of sunlight and darkness that is determined by the rotation of the earth and the position of the sun.) The Egyptians of course gave names to these days and this practice is still continued.

In early calendars a month was the amount of time it took for the Moon to complete an orbit and undergo its cycles. Recall the very early calendars that were established on the cycles of the moon, where each cycle of the moon takes about twenty-nine and a half days, compiling a calendar of twelve months would result in a year that is only three hundred and fifty-four days. The Romans had ten months in their early calendar, ending in the month of "Dece" (ten) and the shortest of years. If we add thirteen months we arrive at three hundred and eighty-three and a half days. There is no correct solution if the moon is the principal inspiration.

Today it is known that it takes three hundred and sixty-five days, five hours, forty-eight minutes, and forty-six seconds for the earth to complete a revolution of the sun, which designates a year. This created the problem of what to do about the extra five hours, and forty-eight minutes, and forty-six seconds. How can the extra time be taken into account? The European answer was "Leap Year." To make up that extra time, one day is added to the calendar every four years. Because of this, many uncertainties have appeared, causing other cultures to retain (or create) their own calendars.

In the Hebrew calendar (as in the Egyptian calendar), the New Year starts in the fall. They have a peculiar system that allows them a twelve-month calendar that in certain years can be extended to include an additional month. This extra month is added seven times over every period of nineteen years, called the Metonic cycle. Each month is either twenty-nine days or thirty days, depending on changes in the moon. The names of the months are Tishri, Heshvan, Kislev, Tebet, Shebat, Adar, Nisan, Iyar, Sivan, Tammuz, Ab, and Elul. The extra month, Veadar, follows Adar.

The Chinese calendar also has twelve lunar months, with a thirteenth added in leap year. In this oriental version the New Year never begins or occurs on the day that was celebrated last year. The new year begins during the month in which the sun enters a certain star constellation (or zodiac) in the sky. This usually occurs between the twentieth of January and the nineteenth of February. Each year is named for one of twelve animals: the Mouse, Ox, Tiger, Rabbit, Dragon, Snake, Horse, Sheep, Monkey, Rooster, Dog, and Pig.

The Islamic or Muslim calendar has twelve months that have either twenty-nine, or thirty days and is in tune with the cycles of the moon also. But extra days or months are not added to keep time with the sun. Thus in, thirty-three Muslim years are equal to thirty-two solar years. The seasons do not keep in steady line with the calendar. New Year's Day, for example, moves year by year (over thirty-three years) through winter, spring, summer, and fall. It would appear that the moon is the staple of all calendars except that which was developed by those ancient Africans. Unfortunately the Inca and the Mayan calendars were also influenced by the phases of the moon.

From the Zodiac, such as the one found at Dendera, the ancient Egyptians formulated the ten-month solar calendar. This understanding is reflected in the "Rasaurian Solar Calendar" that introduces itself as a star chart, derived from the African concept of time and space. It would seem that the calendar can only be the result of celestial movement, be it that of the moon or of the stars.

Consider this premeditation for a calendar. Outer space was the timeless void before the introduction of the light. RA was the light and as the light moved within the void, time was introduced. Our sun is a member of a trinity of suns that played major roles in the development of our quadrant of stars, planets, moons, and life forms. The "word" was given and the three stars began to move apart from one another. Their names were RA, Sothis (Sirrus), and Hathor (Alcyone), As this separation occurred the surface of each was lessened by great masses of viscous matter that seemed to escape into space, only to be captured by the gravitational pull of its inception. These three stars highlight much of what we see today as we gaze upward and outward. Our view is restricted only by the natural alignment of our earth and our northern hemisphere.

From these bodies of light come the makings of our zodiac, our solar system, and all the visible constellations. With this natural order as their grid, the ten-month solar calendar was fashioned. Our solar system has ten heavenly bodies, one sun and nine planets. Thus they factored together a calendar of ten months, each consisting of thirty-six days, four nine-day weeks.

They developed a system of 60 seconds to a minute and 60 minutes to an hour. This was done by adding the 12 miles a second that the sun travels as it revolves around Alcyone, "ten times," that gave them 120 miles of movement in ten seconds; by repetition they arrived at 720 miles of movement in 60 seconds, 1420 in an hour, and 34,080 miles in 24 hours, then realizing that at each cycle of sixty," a new star alignment occurred. Thus the sidereal year was born. It came to be understood that a sidereal sighting of Sothis occurred annually at the same time. The sighting of Sothis provided the answers to the number of hours in one orbit of the earth around the sun, which is 8766 hours. Armed with this information, a calendar could be accurately devised consisting of 10 months, of 360 days, 4 weeks a month, nine days a week, and 8766 hours a year.

What made the Solar calendar unique was the concept of time allotment, such as allotting twelve hours to the day and twelve hours to the night and thirty-five minutes that were split between the "Tuat" and the "Duat." The latter being the periods between sunset and sunrise, respectively. This resulted in the sightings of familiar stars at a predictable and seasonal occurrence. Each day of the year was heralded by the sighting of stars that gave identity to its significance. In this way they related time, space, and the stars. Every second was accounted for and the natural order was observed.

This produced the first of the ancient Egyptian calendars, known as the solar calendar of 360 (24 hours and 35 minutes) days in a year, based on the sighting of stars. This was an understanding that went beyond the moon and allowed one hundred percent accuracy. Did the ancient Egyptians equate the moon with the calendar? Yes! Strangely enough, in their own way they did.

Thuti, the teacher of Ausaures, was known as the God of the Moon. It was under his instigation that the Thutian calendar, consisting of three hundred and sixty five days and a quarter of a day comes into being. Because of the strong effect that the moon displays in other calendars, that are without a doubt derived from this system, the story continues in classic ancient Egyptian pageantry that weaves fact with fiction and theories that involve the movement of stars.

It was after the earth was saved from the "Insects" and the planet was ready for humankind. RA had forbidden the allotment of more time so that mankind could be born on the earth. Only within the existing order of time would life forms be allowed. All the other forms were created and surviving; only mankind need be added to complete the seven species of life that are found on the earth. Seb petitioned Lord Thuti to plead with RA for an allotment of time so that his mate Nut, the Sky Goddess, could give birth to the earth gods. RA was unrelenting.

It was known that the great company of gods would often gather and engage in games of chance. RA was in need of someone to give names to all the creatures that he had allowed to evolve and in this way his works would have meaning. He was beginning to grow old from the billions and billions of years of existence. He found this task, too tedious. A wager was arranged between RA and Thuti, with the winner indulging the other. Thuti would name all the creatures if he should lose. Should RA lose, the thirty-five minutes that existed between the "Tuat" and the "Duat" would be forfeited.

In a wager that is known only to the gods, Lord Thuti was the victor. He claimed his prize, the thirty-five minutes of time that the he gained control of, leaving an exact twenty-four hours in a cycle of ten months, and of four weeks of nine days, totaling 360. The thirty-five minutes when multiplied by 360 totaled five and one quarter days added in a year that was still 8766 hours.

This brings us to another question: How could the ancient Egyptians add five and a quarter days and not have a Leap Year? It also clarifies their final connection with the usage of the moon in their calendar. If we recall that on the Zodiac of Dendera there is mention of the Epagamal days or the ancient Egyptian New Year, these are the days that were gained by Lord Thuti. The "five and one quarter days" that were added on at the end of the 360 earth orbit were the days that the earth gods were born. In the Rasaurian Solar Calendar they are visible in the month of Thuti (with the moon going through its phases), signaling the sighting of the star Sirrus and the "five and one quarter days" of observance, until the stars align again, starting the year anew. These "five and one quarter days" were accounted for by observance and waiting for that star grid to align again. Every second of the orbit was accounted for in that singular revolution. There was no need of Leap Year.

This is the understanding that was lost with the intervention of Julius Caesar and the Romans. It is obvious that the Romans dissected the Thutian calendar that is used today, adding Latin names and interpretations and the concept of "Leap Year." The word *calendar* originates from the Latin word *kalendae*.

This Julian calendar advanced the European concept of seasonal change but it was not perfect. We have already mentioned that in the Middle Ages (1582) Pope Gregory XIII made changes to account for the seasonal discrepancies. For one year he dropped ten days from the calendar to make it correspond more closely with the seasons. He also dropped leap years in "century" years, unless those years could be divided by four hundred. This gap in generic understanding created confusion and his changes were rejected. Today is it known as the Gregorian or New Style calendar.

In a generic star-dated calendar the days of the month do not move forward or backwards. They are constant, and each is identified by the sighting of a star the evening before. In this new-style calendar, one's day of birth moves from year to year; the cycle repeats itself after seven years. This results in failure to find stars that were present in the sky at the time of your actual birth and the confusion about identifying the name of that day. The Rasaurian Solar Calendar is all of these calendars in one, with three versions that encompass the message that was depicted on the ceiling zodiac in the Egyptian temple of Dendera.

In closing, let us reflect on the two laws of existence. The first natural law of all living creatures is to exist. The second natural law for all living creatures is to co-exist together.

The End

Appendix A List of Illustrations
(ArtWork by Dana Williams)

Appendix B GLOSSARY

Apepl The serpent of darkness, aka the Devil

Apex of the sun's way A point toward which the sun is traveling, carrying with it the entire solar system

"Aquarian Age" The age of comprehension and earthly brotherhood, the end of hunger, war, and man's indifference to his fellow man (the end of racism)

C.M.B. Cosmic microwave background anisotropy, tiny fluctuations in the sky's brightness at a level of.000001 or 1 part in one hundred thousand or 1/100,000

COBE Cosmic observatory background explorer, launched Nov. 18, 1989, and designed to measure the diffuse infrared and microwave radiation

Disorder of Existence Big Bang theories and others that suggest our being and creation was happenstance and had no pre-thought

DMR Differential microwave radiometer, used by the COBE to detect for the first time and characterize faint fluctuations in the CMB

Egyptian concepts Expressions of modern scientific truths

FIRAS Far Infra Red Absolute Spectrophotometer, made the precise measurement of the spectrum of the CMB

Hakka Power, magic, the self-created one

Humankind The highest form or level of the seven (7) species of earth life

Intrafarance The ability of particle interaction affecting the heat and the movement of "Light" and "Radiation"

Mankind The measure of the Universe

Mayhem The Solar Serpent

Movement of Translation The Sun (RA) has an actual movement (and the other Stars are also in motion as well) this movement is at a rate of twelve (12) miles + or - a second toward the constellation Cygnus in the general direction of the brilliant Star Vega

NUT The Goddess of the Sky

Order of existence Issued for the survival of all things and can be followed back to the concept of predefinitive consciousness that had a comprehensive plan for the creation and re-creation that is autonomous

Pelades A small constellation of seven (7) stars with a central Star called Alcyone, which the other stars orbit directly within this constellation. Observing this star system from within the inner Milky Way and looking outward, there would be an eighth star which would be the real completion of this system. This star is our sun (RA) and our complete solar system

Pert The symbol of the sky

Phenomenon Reactions of living substances to radiant energy (photosynthesis)

Pscean Age Starting at 666 B.C., a 2,600-year cycle which marked the emergence of the European civilization

Precession The earth rotating with a slow westward shift of the equinoctial points along the plane of the ecliptic (similar to a spinning top as it loses speed. It causes the polar axis to indicate a different point in space each way. Each cycle is 2,600 years "an Age," working the astrological chart in reverse

Racism The idea that there is no Universal plan for mankind (the absence of light within oneself)

RA Our sun, which is 81 percent hydrogen, 18 percent helium, and 1 percent oxygen, magnesium, nitrogen, and other elements (what are solid on the earth are gases within the sun)

Ranebertecher The Lord of the Light

SEB God of Solid

SHU God of Gas

Star dust Mankind

Upuat The Opener of Ways

Urgeist Primeval spirit, according to Dr. Henrich Brugsh, that felt the desire for creative activity and thought. Its "Words" Awoke the world to life and that this first act of creation began the formation of the primeval watery mass of an "egg"

Walter A. Fairservis Jr Author of *The Ancient Kingdoms of the Nile* where he confirms archaeological evidence that life most likely existed on the shores of the Nile river two million years (2,000,000) ago. Those inhabitants of the Nile valley believed in the Creator of the Light and the afterlife, thereby establishing an African religious theory that predates any commonly believed historical supposition.

Appendix C List Of References

1.) **E.A. Wallis Budge:**

   The Book of the Dead
   The Gods of the Egyptians Vol. I and II

2.) **Raschwaller De Lubicz:**

   Sacred Science
   Her-Bak Vol. I and II

3.) **Rudolpho Benavides**

   Dramatic Prophecies of the Great Pyramid

4.) **Evan Hadingham**

   Early Man and the Cosmos

5.) **The Old Encyclopedia Brittanica**

6.) **The Book of Knowledge "Childrens Encyclopedia"**

   The New Book of Knowledge

7.) **Walter Fairseruis Jr.**

   The Ancient Kingdoms of the Nile

8.) **Dr. Henrich Brugsh**

   Religion and Mythology Der Alten Aegypt

9.) **R.B. Wallace**

   The Old Religion of Ancient Egypt

10.) **The Holy Bible**

   Book of Daniel

11.) **N.A.S.A.**

The Gamma Ring Primeval Germ and Intrafarance

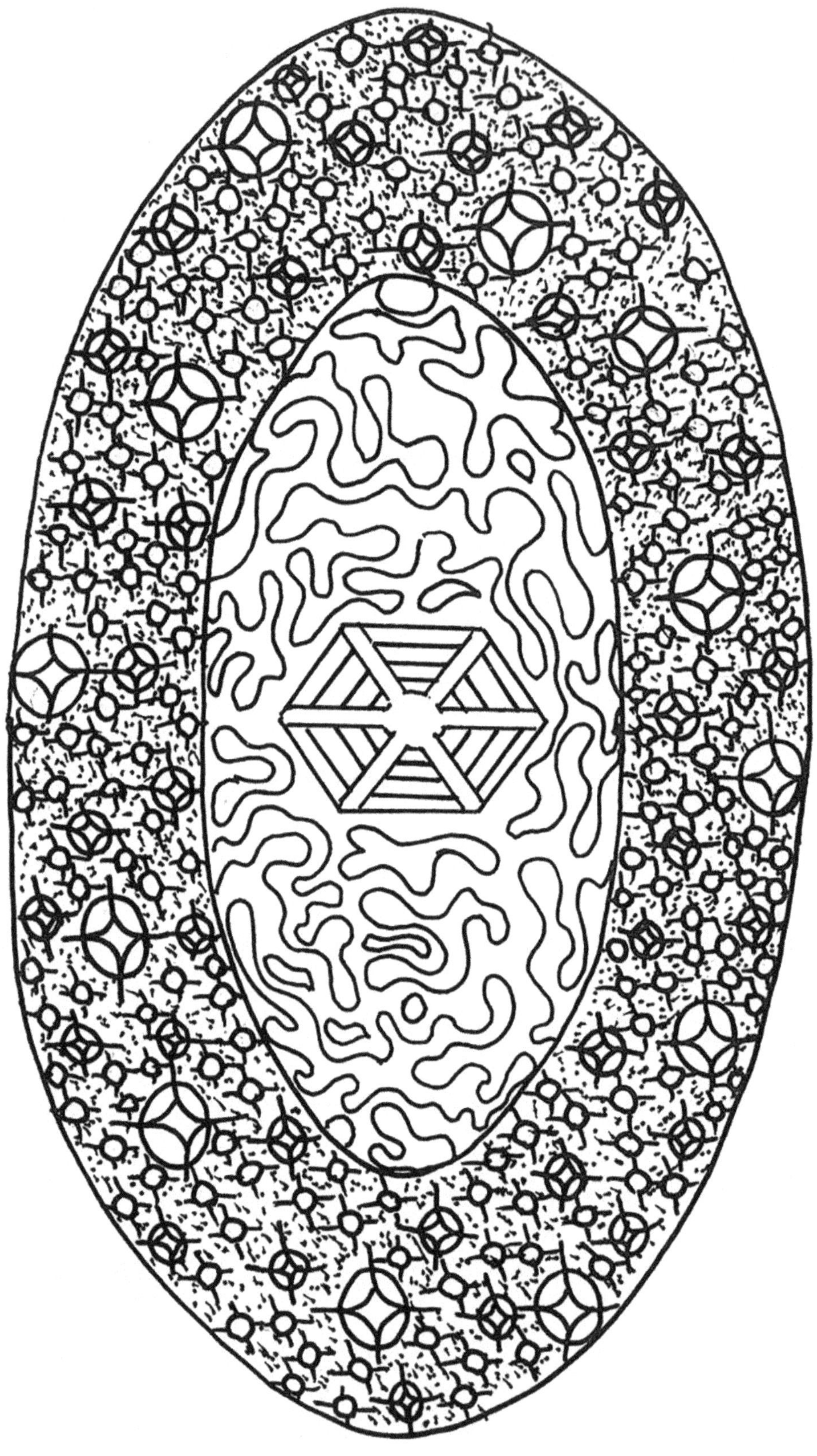

The Universe Black Holes/Gamma Ring/Primeval Germ

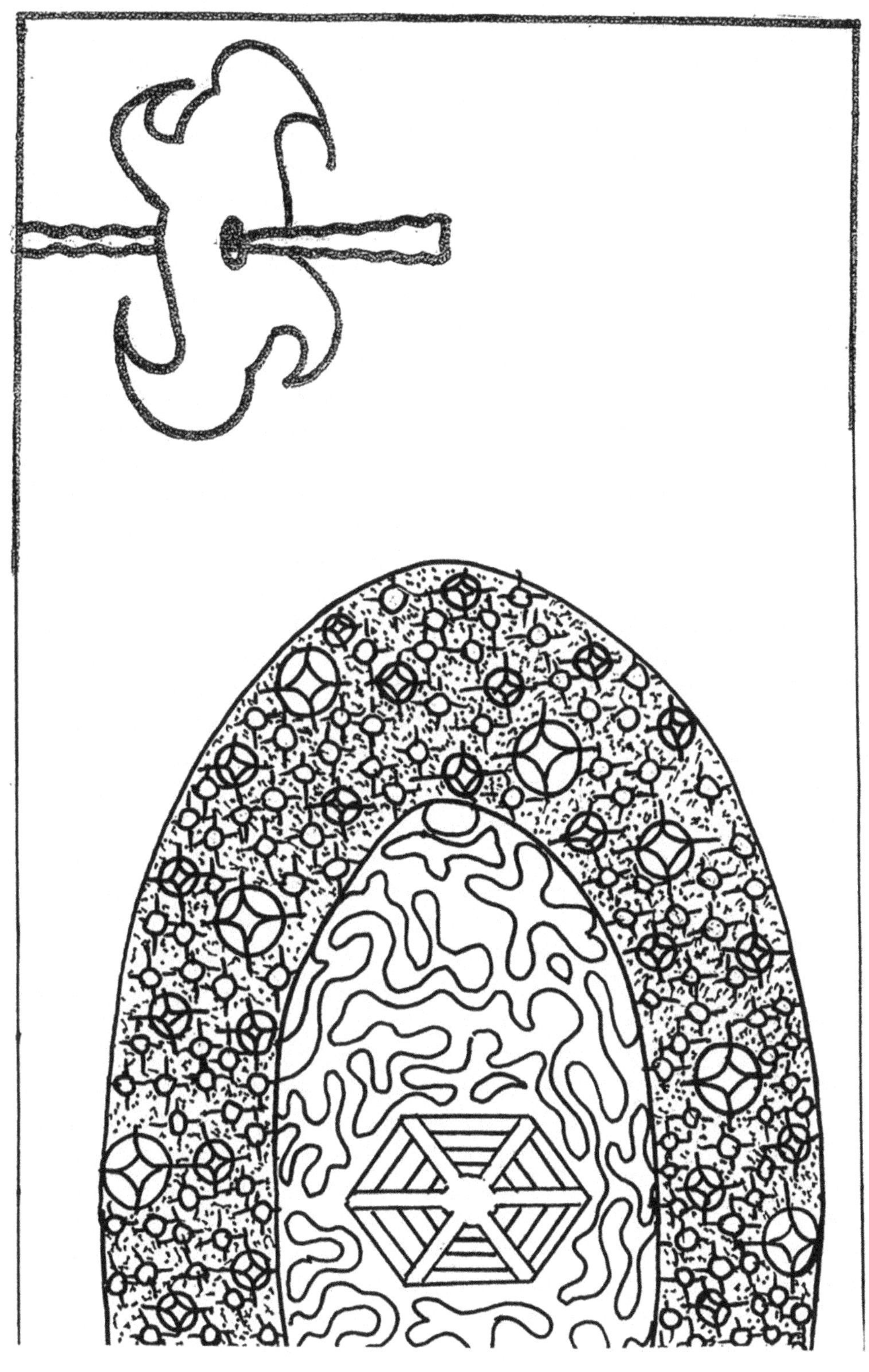

The universe of light/blackholes/gamma belt/primeval germ

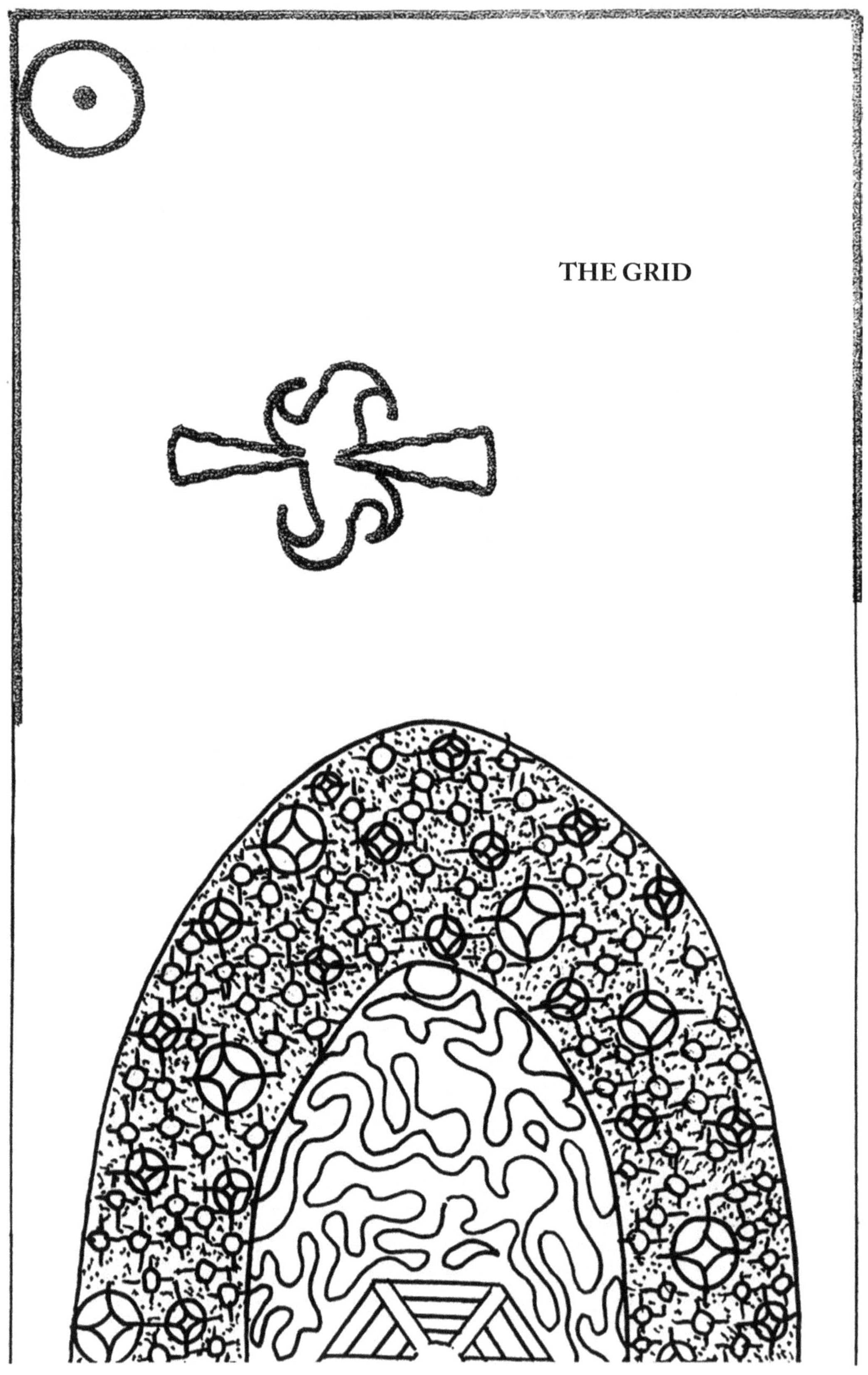

THE CIRCLE

If Time Travel was possible, light travelling in circles may create repition of pattern.
We need an assimalator a capacitor or accumulator

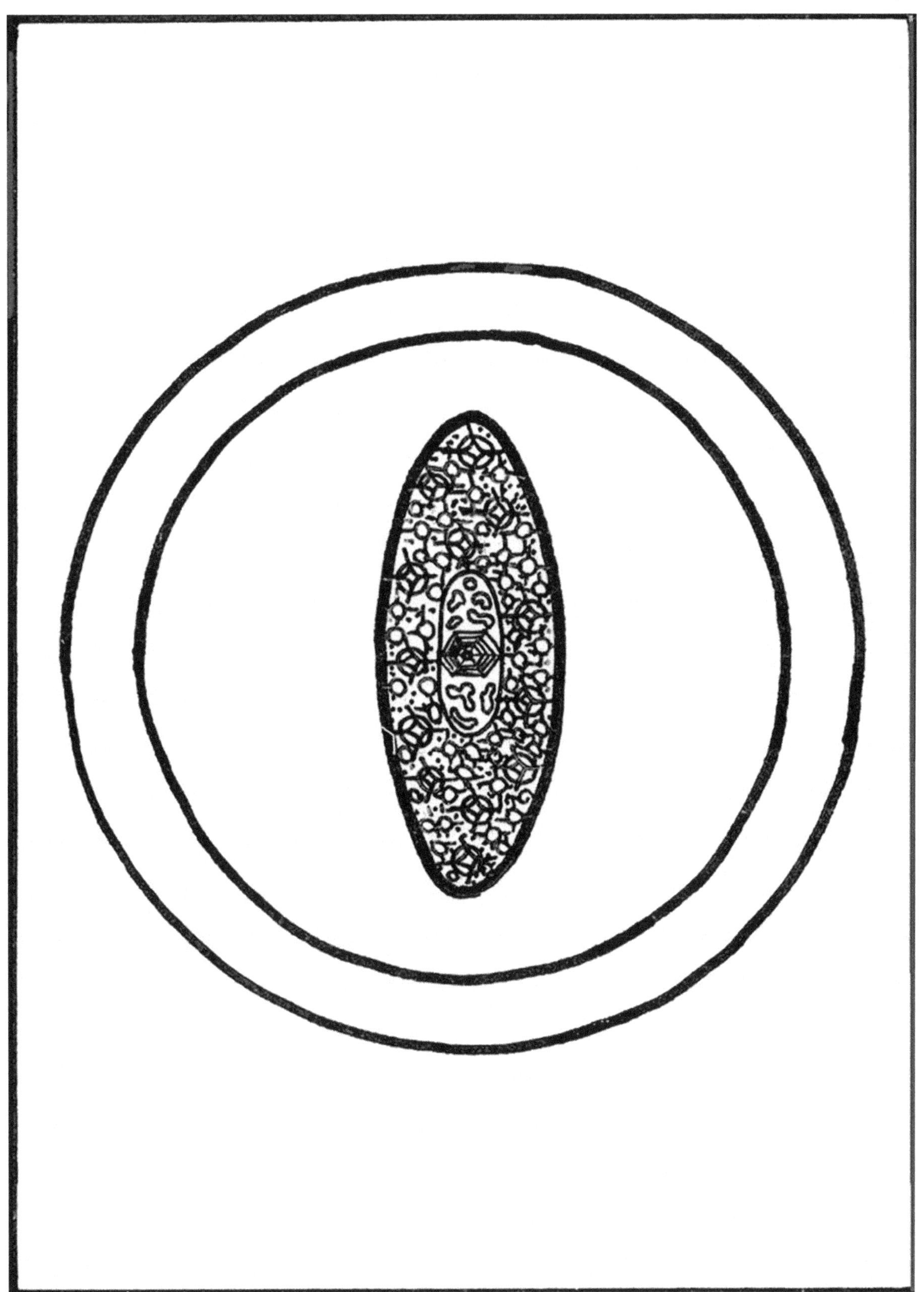

THE TRIANGLE

This shape is a natural resistor to the movement of extended thought.

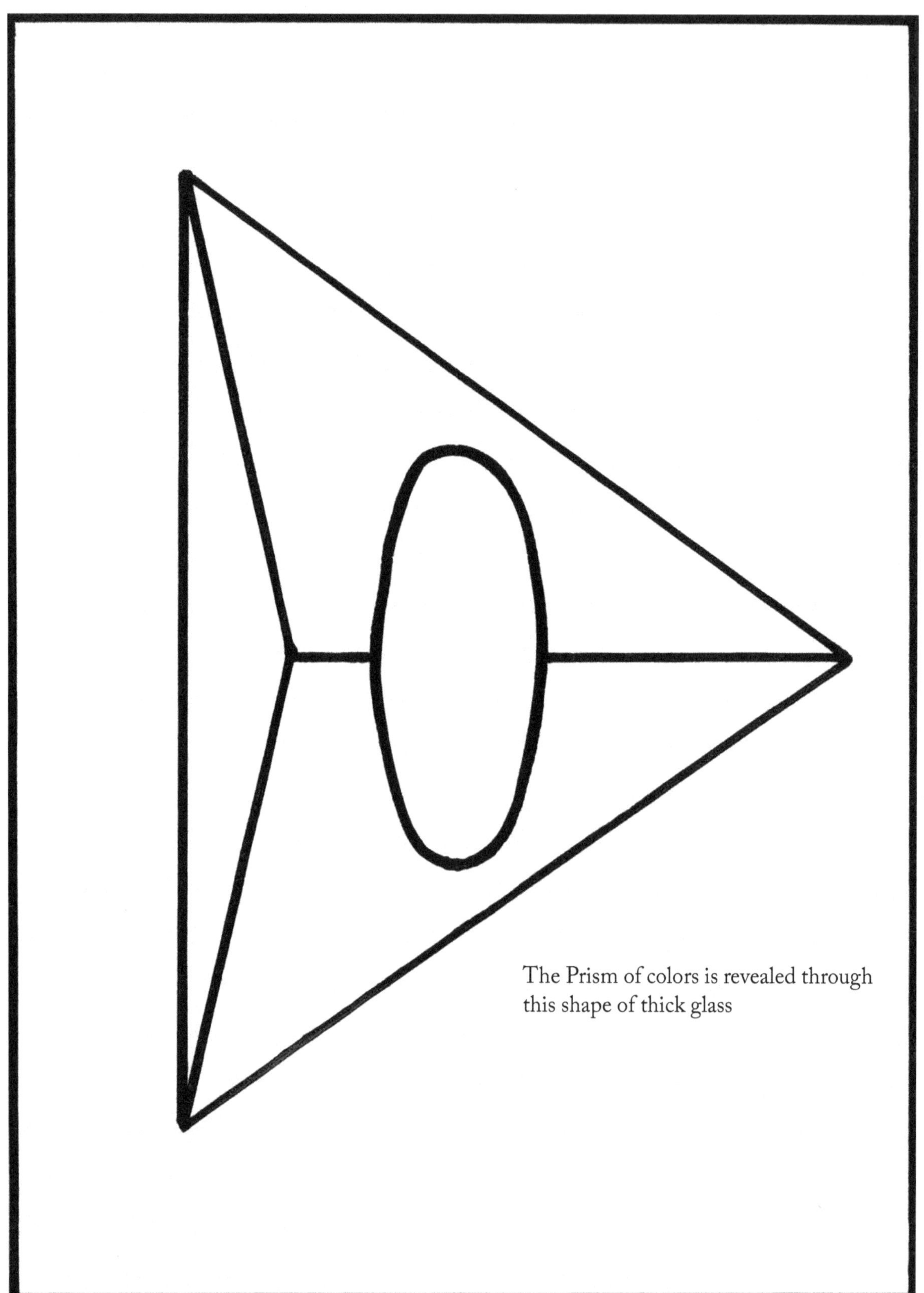

THE RASAURIAN MODEL

The CUBE allows energy containment. It suggest inner and outer with beginnings
and infinite ends.

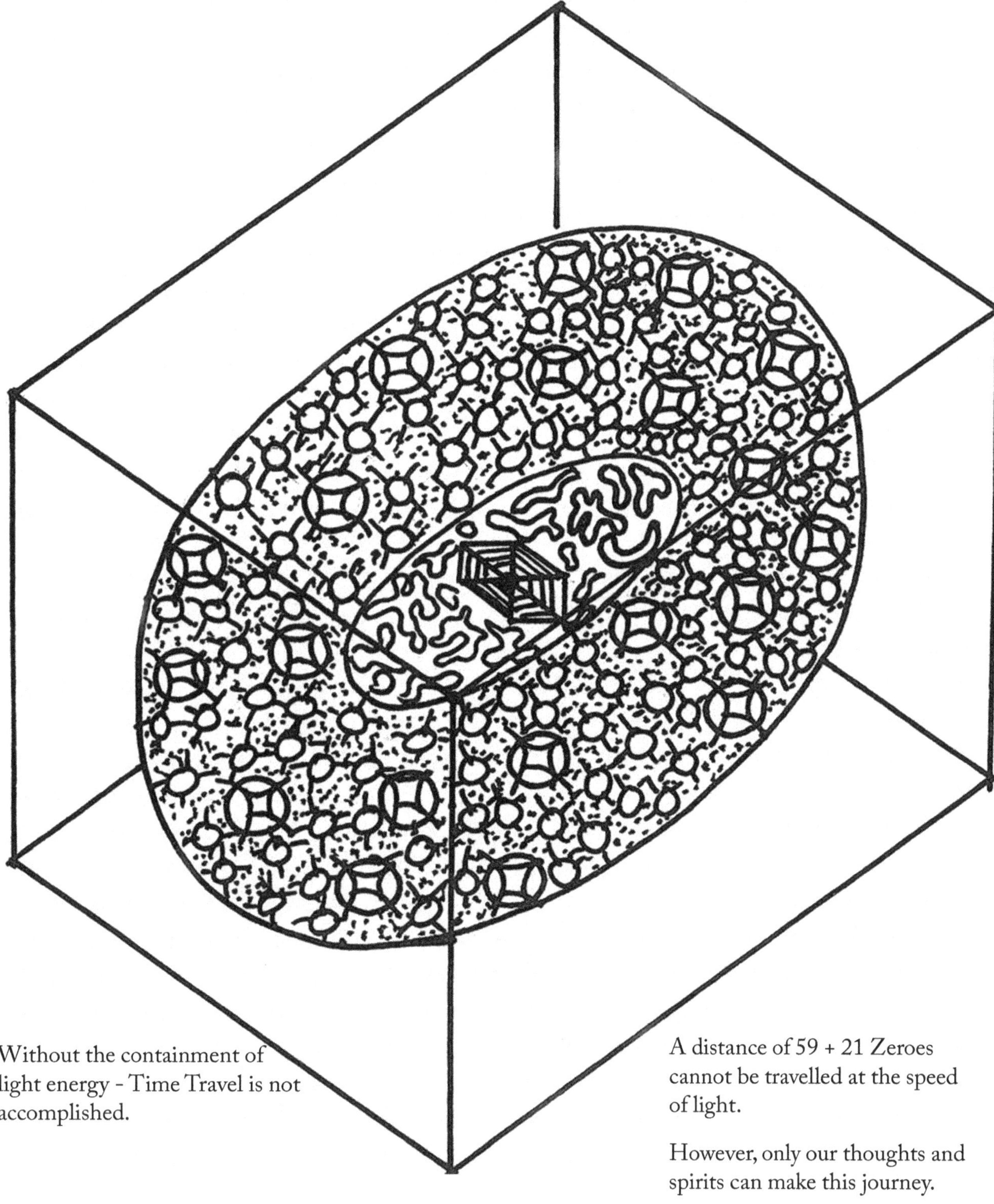

Without the containment of
light energy - Time Travel is not
accomplished.

A distance of 59 + 21 Zeroes
cannot be travelled at the speed
of light.

However, only our thoughts and
spirits can make this journey.

The Turtle represents time-les-ness. Before the beginning of time. He is centered between the symbols of "Energy Everlasting" and "Intelligence that is uncomprehendable" indications that which is "Eternal"